Energy Audits

Energy Audits

A Workbook for Energy Management in Buildings

Tarik Al-Shemmeri
Professor of Renewable Energy Technology
Staffordshire University

WILEY-BLACKWELL

A John Wiley & Sons, Ltd., Publication

Library of Congress Cataloging-in-Publication Data

Al-Shemmeri, Tarik.
    Energy audits : a workbook for energy management in buildings / Tarik Al-Shemmeri.
        p. cm.
    Includes bibliographical references and index.
    ISBN-13: 978-0-470-65608-2 (pbk.)
    ISBN-10: 0-470-65608-5 (pbk.)
    1. Buildings–Energy conservation–Handbooks, manuals, etc.   2. Energy auditing–Handbooks, manuals, etc.   3. Energy conservation–Handbooks, manuals, etc. I. Title.
    TJ163.3.A35 2011
    658.26–dc23

                                                    2011021991

A catalogue record for this book is available from the British Library.

This book is published in the following electronic formats: ePDF 9781119950295; ePub 9781119950301; Mobi 9781119950318

Set in 9/12.5pt Interstate by SPi Publisher Services, Pondicherry, India
Printed in Singapore by Ho Printing Singapore Pte Ltd

1   2011

Contents

Preface

This book is concerned with the subject of energy: as a resource, its use and the ways in which it can be optimised.

The need for this type of book is very well documented, and below I will try to articulate that.

Coal fuelled the Industrial Revolution in the 18th and 19th centuries. It remained as the prime fuel until well into the 20th century, supplying the steam engines used in road vehicles and trains. The industrialised nations were in huge competition in their search for additional fuels and the discovery of oil in the Middle East and elsewhere extended the use of this fuel, opening up a broader scope for applications. With the advent of the automobile, aeroplanes and the spreading use of electricity, oil became the dominant fuel during the 20th century. During the Arab-Israeli wars in 1967 and 1972, the price of oil increased from 5 to 45 US dollars per barrel and there was a resultant shift away from oil. Coal and nuclear became the fuels of choice for electricity generation, and conservation measures increased energy efficiency. The use of fossil fuels has continued to grow and their share of the energy supply has increased. The more recent Gulf Wars in the 1990s and the new millennium, and the accompanying invasion of Iraq are, some would say, clear evidence of the importance of oil to the West.

In 2005, total worldwide energy consumption was 500 EJ (= 5×10^{20} J or 138 900TWh) with 86.5% derived from the combustion of fossil fuels. Estimates of the remaining worldwide energy resource vary, with the remaining fossil fuels totalling an estimated 0.4 YJ (1 YJ = 10^{24} J).

Political considerations over the security of supplies and environmental concerns related to global warming and sustainability will move the world's energy consumption away from fossil fuels.

In order to drive the move away from fossil fuels, we are likely to witness further examples of economic pressure through such measures as carbon trading and green taxation. Some countries are taking action as a result of the Kyoto Protocol, and further steps in this direction are proposed. For example, the European Union Commission has proposed that the Energy Policy should set a binding target of increasing the level of renewable energy in the EU's overall mix from less than 7% today to 20% by 2020.

This book tackles the fundamental principles of thermodynamics in day-to-day engineering concepts; it provides the tools to measure process efficiency and sustainability accurately in power and heating applications - helping engineers to recognise why losses occur and how they can be reduced by utilising familiar thermodynamic principles.

Acknowledgements

I would like to express my appreciation and gratitude to the many people who have put up with me through the period of writing this book; to all those who offered support, read, wrote, offered comments, to those who assisted in the editing, proofreading and design.

I wish to thank all of my students who have been with me over the course of the years and helped me develop the material into its present form.

I would like to thank my family and friends and especially my son Mark who helped me creating most of the illustrations and graphs in this book.

Grateful thanks are due to those who allowed me to present their case study material in the various chapters of this book.

Last but not least, I beg the forgiveness of all others who have helped me but I have failed to mention here.

Dimensions and Units

Any physical situation, whether it involves a single object or a complete system, can be described in terms of a number of recognisable properties which the object or system possesses. For example, a moving object could be described in terms of its mass, length, area or volume, velocity and acceleration. Its temperature or electrical properties might also be of interest, while other properties – such as the density and viscosity of the medium through which it moves – would also be of importance, since they would affect its motion. These measurable properties used to describe the physical state of the body or system are known as its variables; some variables are basic, such as length and time, while others are derived, such as velocity. Each variable has units to describe the magnitude of that quantity. Lengths in SI units are described in units of metres. The metre is the unit of the dimension of length (L); hence, the area will have dimension L^2, and volume L^3. Time (t) will have units of seconds; hence, velocity is a derived quantity with dimensions of (Lt^{-1}) and units of metres per second (m/s or ms^{-1}). A list of some variables together with their units and dimensions is given in Table 1, while further examples of units and conversion factors are given in Tables 2 to 7.

Definitions of some basic SI units

Mass: The kilogram is the mass of a platinum-iridium cylinder kept at Sèvres in France.

Length: The metre is the distance between two engraved lines on a platinum-iridium bar also kept at Sèvres. The measurement is made at 0°C and standard atmospheric pressure, the bar being supported in a specified manner. The metre is now defined as being equal to 1 650 763.73 wavelengths in a vacuum of the orange line emitted by the Krypton-86 atom.

Time: The second is defined as the fraction 1/31 556 925.975 of the tropical year for 1900. The second is also declared to be the interval occupied by 9 192 631 770 cycles of the radiation of the caesium atom corresponding to the transition between two closely spaced ground state energy levels.

Temperature: The Kelvin is the degree interval on the thermodynamic scale on which the temperature of the triple point of water is 273.16 K exactly. (The temperature of the ice point is 273.15 K.)

Definitions of some derived SI units

Force: The Newton is that force which, when acting on a mass of one kilogram gives it an acceleration of one metre per second per second.

Work energy, Quantity of heat: The joule is the work done by a force of one Newton when its point of application is moved through a distance of one metre in the direction of the force. The same unit is used for the measurement of every kind of energy including quantity of heat. The Newton metre, the joule and the watt second are identical in value. It is recommended that the Newton is kept for the measurement of torque or moment and the joule or watt second is used for quantities of work or energy.

Table 1   Basic SI units.

Quantity	Unit	Symbol
Length [L]	Metre	M
Mass [m]	Kilogram	Kg
Time [t]	Second	S
Electric current [I]	Ampere	A
Temperature [T]	Degree Kelvin	K
Luminous intensity [Iv]	Candela	Cd

Table 2   Derived units with special names.

Quantity	Unit	Symbol	Derivation
Force [F]	Newton	N	$kg\text{-}m/s^2$
Work, energy [E]	Joule	J	N-m
Power [P]	Watt	W	J/s
Pressure [p]	Pascal	Pa	N/m^2

Table 3 Some examples of other derived SI units.

Quantity	Symbol
Area	m^2
Volume	m^3
Density	kg/m^3
Angular acceleration	rad/s^2
Velocity	m/s
Pressure, stress	N/m^2
Kinematic viscosity	m^2/s
Dynamic viscosity	$N\text{-}s/m^2$
Momentum	$kg\text{-}m/s$
Kinetic energy	$kg\text{-}m^2/s^2$
Specific enthalpy	J/kg
Specific entropy	J/kgK

Table 4 Non-SI units.

Quantity	Unit	Symbol	Derivation
Time	Minute	min	60s
Time	Hour	h	3.6 ks
Temperature	Degree Celsius	°C	K − 273.15
Angle	Degree	°	$\pi/180$ rad
Volume	Litre	L	$10^{-3}\,m^3$ or dm^3
Speed	Kilometre per hour	km/h	–
Angular speed	Revolution per minute	rev/min	–
Frequency	Hertz	Hz	cycle/s
Pressure	Bar	b	$10^2\,kN/m^2$
Kinematic viscosity	Stoke	St	$100\,mm^2/s$
Dynamic viscosity	Poise	P	$100\,mN\text{-}s/m^2$

Table 5 Multiples of units.

Name	Symbol	Factor	Number
exa	E	10^{18}	1 000 000 000 000 000 000
peta	P	10^{15}	1 000 000 000 000 000
tera	T	10^{12}	1 000 000 000 000
giga	G	10^9	1 000 000 000
mega	M	10^6	1 000 000
kilo	k	10^3	1000
hecto	h	10^2	100
deca	da	10	10
deci	d	10^{-1}	0.1
centi	c	10^{-2}	0.01
milli	m	10^{-3}	0.001
micro	μ	10^{-6}	0.000001
nano	n	10^{-9}	0.000000001
pico	p	10^{-12}	0.000000000001
fempto	f	10^{-5}	0.000000000000001
atto	a	10^{-18}	0.000000000000000001

Table 6   Conversion factors.

Item	Conversion
Length	1 in = 25.4 mm
	1 ft = 0.3048 m
	1 yd = 0.9144 m
	1 mile = 1.609 km
Mass	1 lb = 0.4536 kg (0.453 592 37 exactly)
Area	1 in² = 645.2 mm²
	1 ft² = 0.092 90 m²
	1 yd² = 0.8361 m²
	1 acre = 4047 m²
	1 mile² = 2.590 km²
Volume	1 in³ = 16.39 cm³
	1 ft³ = 0.028 32 m³ = 28.32 litres
	1 yd³ = 0.7646 m³ = 764.6 litres
	1 UK gallon = 4.546 litres
	1 US gallon = 3.785 litres
Force, Weight	1 lbf = 4.448 N
Density	1 lb/ft³ = 16.02 kg/m³
Velocity	1 km/h = 0.2778 m/s
	1 ft/s = 0.3048 m/s
	1 mile/h = 0.4470 m/s = 1.609 km/h
Pressure, Stress	1000 Pa = 1000 N/m² = 0.01 bar
	1 in H₂O = 2.491 mb
	1 lbf/in² (Psi) = 68.95 mb or 1 bar = 14.7 Psi
Power	1 horsepower = 745.7 W
Moment, Torque	1 ft-pdl = 42.14 mN-m
Rates of flow	1 gal/h = 1.263 ml/s = 4.546 L/h
	1 ft³/s = 28.32 L/s
Fuel consumption	1 mile/gal = 0.3540 km/L
Kinematic viscosity	1 ft²/s = 929.0 cm²/s = 929.0 St
Dynamic viscosity	1 lbf-s/ft² = 47.88 N-s/m² = 478.8P
	1 pdl-s/ft² = 1.488 N-s/m² = 14.88P
	1 cP = 1 mN-s/m²
Energy	1 horsepower-h = 2.685 MJ
	1 kW-h = 3.6 MJ
	1 Btu = 1.055 kJ
	1 therm = 105.5 MJ

Table 7 Conversion factors.
Simply multiply the imperial by a constant factor to convert into metric or the other way around.

	Unit	× Factor	= Unit	× Factor	= Unit
Length (L)	ins	25.4	mm	0.0394	ins
	ft	0.305	m	3.281	ft
Area (A)	in²	645.16	mm²	0.0016	in²
	ft²	0.093	m²	10.76	ft²
Volume (V)	in³	16.387	mm³	0.000061	in³
	ft³	0.0283	m³	35.31	ft³
	ft³	28.32	litre	0.0353	ft³
	pints	0.5682	litre	1.7598	pints
	Imp. gal	4.546	litre	0.22	Imp gal
	Imp. gal	0.0045	m³	220	Imp gal
Mass (M)	lb	0.4536	kg	2.2046	lb.
	tonne	1000	kg		
Force (F)	lb	4.448	N	0.2248	lb.
Velocity (V)	ft/min	0.0051	m/sec	196.85	ft/min
Volume flow	Imp gal/ min	0.0758	litres/s	13.2	Imp gal/ min
	Imp gal/h	0.00013	m³/s	7,936.5	Imp gal/h
	ft³/min	0.00047	m³/s	2,118.6	ft³/min
Pressure (P)	lb/in²	0.0689	bar	14.5	lb/in²
	kg/cm²	0.9807	bar	1.02	kg/cm²
Density (ρ)	Lb/ft³	16.019	kg/m³	0.0624	lb/ft³
Heat flow rate (Q)	Btu/h	0.2931	W	3.412	Btu/h
	kcal/h	1.163	W	0.8598	kcal/h
Thermal conductivity (k)	Btu/ft h R	1.731	W/mK	0.5777	Btu/ft h R
	kcal/m h K	1.163	W/mK	0.8598	kcal/m h K
Thermal conductance (U)	Btu/h ft² R	5.678	W/m²K	0.1761	Btu/h ft² R
	kcal/h m² K	1.163	W/m²K	0.8598	kcal/h m² K
Enthalpy (h)	Btu/lb.	2,326	J/kg	0.00043	Btu/lb.
	kcal/kg	4,187	J/kg	0.00024	kcal/kg

List of Figures

List of Tables

Chapter 1

Energy and the Environment

Learning outcomes

• Demonstrate the various forms of energy: mechanical, electrical, chemical and nuclear	Knowledge and understanding
• Evaluate the amount of energy for various types	Analysis
• Solve problems associated with energy conversion	Problem solving
• Distinguish between: potential and kinetic energy, electrostatic and electromagnetic energy, chemical and thermal energy, energy from nuclear fission and nuclear fusion	Knowledge and understanding
• Calculate the calorific value of fuels	Analysis
• Calculate the combustion products of fuels	Analysis
• Examine future world energy scenarios	Reflections
• Practise further tutorial problems	Problem solving

Energy Audits: A Workbook for Energy Management in Buildings, First Edition.
Tarik Al-Shemmeri.
© 2011 Blackwell Publishing Ltd. Published 2011 by Blackwell Publishing Ltd.

1.1 Introduction

It is necessary to appreciate that energy will be needed to modify the state of any working environment and keep the conditions comfortable, whether by providing warm or cool air.

Heating the air is a simple process of increasing its thermal energy and consequently raising its temperature. Heating of air can be achieved either by direct or indirect means: examples such as coal and gas fires represent the direct forms of heating normally used in domestic situations; electrical resistance heating elements provide an example of an indirect method because electricity is produced elsewhere. Hot water radiators are also used to provide indirect heating of indoor air.

Similarly, the process of cooling air, or reducing its thermal energy, is energy driven.

Both cooling and heating require a further process. After being treated (i.e. heated or cooled), the air has to be delivered to the space where it is needed, and hence an electric fan is usually used to push the air from the apparatus out into the room.

The following sections will discuss the various forms of energy, and how energy can be converted from one form to another form which is convenient for heating, cooling, etc.

This chapter will demonstrate the environmental impact of using different fuels to provide energy for heating or electricity.

1.2 Forms of energy

We associate energy with devices whose inputs are fuel based, such as electrical current, coal, oil or natural gas, and whose outputs involve movement, heat or light.

The unit of energy is the Joule (J). The rate of producing energy is called *power*, and this has the unit Joules per second (Js^{-1}) or the Watt (W).

There are five forms of energy:

- mechanical energy;
- electrical energy;
- chemical energy;
- nuclear energy;
- thermal energy.

1.2.1 Mechanical energy

This type of energy is associated with the ability to perform physical work. There are two forms in which this energy is found; namely *potential energy* and *kinetic energy*.

Potential energy

This is energy contained in a body due to its height above its surroundings. An example is the gravitational energy of the water behind a dam.

Potential energy = mass × acceleration due to gravity (9.81) × height above datum

Or

$$E_p = m \times g \times h \qquad\qquad [1.1]$$

The energy produced by one kilogram of water falling from a height of 100 m above ground is an example of potential energy, and can be calculated as follows:

Potential energy = mass × acceleration due to gravity × height above datum

$$E_p = 1 \times 9.81 \times 100 = 981\,\text{J/kg}$$

Kinetic energy

Kinetic energy is related to the movement of a particular body. Examples of kinetic energy are the flywheel effect and the energy of water flowing in a stream.

Kinetic energy = ½ mass × velocity squared

Or

$$E_k = \tfrac{1}{2} \times m \times v^2 \qquad\qquad [1.2]$$

The water in a river flowing at a velocity of 2 m/s has a kinetic energy of:

Kinetic energy = ½ mass × velocity squared = $\tfrac{1}{2} \times 1 \times (2)^2 = 2\,\text{J/kg}$

1.2.2 Electrical energy

This type of energy, as the name implies, is associated with the electrons of materials. Electrical energy exists in two forms: *electrostatic energy* and *electromagnetic energy*.

Electrostatic energy

This type of electrical energy is produced by the accumulation of charge on the plates of a capacitor. Charles Coulomb first described electric field strengths in the 1780s. He found that for point charges, the electrical force varies directly with the product of the charges; the greater the charges, the stronger the field. And the field varies inversely with the square of the distance between the charges. This means that the greater the distance, the weaker the force becomes. The formula for electrostatic force, F, is given by:

$$F = k\,(q_1 \times q_2)/d^2 \tag{1.3}$$

where q_1 and q_2 are the charges and d is the distance between the charges. k is the proportionality constant, which depends on the material separating the charges.

Electromagnetic energy

This type of energy is produced with a combination of magnetic and electric forces. It exists as a continuous spectrum of radiation. The most useful type of electromagnetic energy comes in the form of solar radiation transmitted by the sun, which forms the basis of all terrestrial life.

1.2.3 Chemical energy

This is associated with the release of thermal energy due to a chemical reaction between certain substances and oxygen. Burning wood, coal or gas, for example, is the main source of energy we use in heating and cooking.

Calculation of chemical energy

The energy liberated from the combustion of a given mass of fuel with a known calorific value in a combustion chamber of known efficiency is given by:

$$\text{Chemical energy} = \text{mass of fuel} \times \text{calorific value} \times \text{efficiency of combustion} \tag{1.4}$$

Typical coal has an energy value of 26 MJ/kg (refer to Table 1.3), which implies that during the combustion of one kilogram of coal, there will be a release of 26 mega joules of thermal energy.

The energy contained in the food we consume is another example of chemical energy.

Analyses of thermal energy liberated from stored chemical energy during the combustion of coal, oil and natural gas will be discussed later in this chapter.

1.2.4 Nuclear energy

This energy is stored in the nucleus of matter, and is released as a result of interactions within the atomic nucleus.

There are three nuclear reactions: *radioactive decay*, *fission* and *fusion*.

Radioactive decay

Here, one unstable nucleus (radioisotope) decays into a more stable configuration, resulting in the release of matter and energy.

Fission

A heavy nucleus absorbs a neutron, splitting it into two or more nuclei and thereby releasing energy. Uranium U235 has the ability to produce 70×10^9 J/kg.

Einstein proposed the following equation to calculate the energy produced from nuclear fission (i.e. the conversion of matter (m) into energy (E) is related to the speed of light (C)):

$$E = mC^2 \tag{1.5}$$

This reaction forms the basis for current nuclear power generation plants.

Fusion

Two light nuclei combine to produce a more stable configuration and this is accompanied by the release of energy. A heavy water (Deuterium) fusion reaction may produce energy at the rate of 0.35×10^{12} J/kg.

This reaction is yet to be used to produce electricity on a commercial basis.

1.2.5 Thermal energy

Thermal energy is associated with intermolecular vibration, resulting in heat and a temperature rise above that of the surroundings. Thermal energy is calculated for two different regimes.

When the substance is in a pure phase, say if it is a liquid, gas or solid, then:

$$\text{Thermal energy} = \text{mass} \times \text{specific heat capacity} \times \text{temperature difference} \tag{1.6}$$

During a change of phase, such as evaporation or condensation, it can be calculated by:

$$\text{Thermal energy} = \text{mass} \times \text{latent heat} \tag{1.6a}$$

However, if there is a change of phase, say during the condensation of water vapour into liquid, there is an additional amount of heat released while the temperature remains constant during the change of phase. For 1kg of water to be heated at ambient pressure from 20 to 120°C, the requirement is:

$$\text{Thermal energy} = \text{heating water (20-100)}°\text{C} + \text{evaporation at 100}°\text{C}$$
$$+ \text{super-heating vapour (100-120)}°\text{C}$$

Thermal energy $= 1 \times 4.219 \times (100{-}20) + 1 \times 2256.7 + 1 \times 2.01 \times (120{-}100)$
$$= 337.52 + 2256.7 + 40.2$$
$$= 2634.42\,kJ$$

Note that the specific heat capacity for water at 1 atmosphere is:

For liquid state $Cp_f = 4.219\,kJ/kg$
For vapour state $Cp_g = 2.010\,kJ/kg$

The latent heat of evaporation is 2256.7 kJ/kg

1.3 Energy conversion

It is important to understand that losses are encountered when energy is transformed during the various conversions into the final form for a given application. For example, consider a power station using coal as a fuel; the following conversions take place.

Coal is mined, extracted and transported to the site. It is then turned into a fine powder and is fed to the combustion system. Energy is liberated by the combustion of coal (chemical to thermal); in other words, not all coal is burnt completely. The thermal energy of the combustion gases is used to raise steam (thermal to thermal); some thermal energy is lost as the exhaust gases leaving the boiler have a temperature above that of the ambient air outside the boiler. High pressure and temperature steam turns the steam turbine (thermal to mechanical). Not all of the energy of the steam is converted into rotational energy, as the steam leaving the turbine has a temperature well above the ambient. The steam turbine drives the electrical generator (mechanical to electrical). Some losses are dissipated through the mechanical connections between the turbine and the electrical generator. Electricity is used by customers for lighting/heating or to operate electrical devices such as radios, televisions, etc. Electrical devices are designed to operate in an optimum condition; the efficiency of the operation will vary depending on its use, age and maintenance.

The energy input to a coal-fired power station (Figure 1.1) can be analysed in a simple way by considering the energy flow diagram; it may look like this:

100 units of energy are stored in the coal fuel
95 units are converted into thermal energy entering the boiler
70 units are absorbed by the water converting into steam
45 units are converted into mechanical energy at the steam turbine rotor
38 units are converted into electrical energy in the electrical generator
30 units go up to waste through the chimney stack.

It is very important to state that, although the initial part of the system above does not indicate any penalties associated with the fuel preparation, this

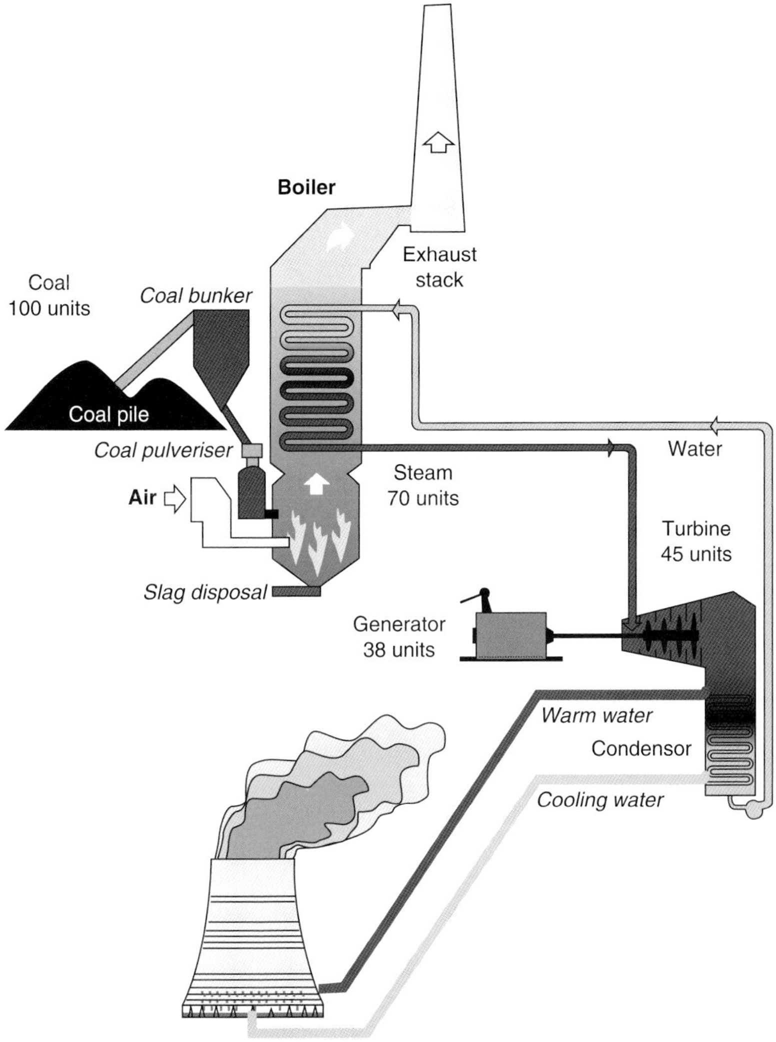

Figure 1.1 Energy conversions in a typical coal-fired power plant (numbers quoted represent the units of energy; dark grey represents warm water and pale grey represents cooling water).

aspect is of particular importance, especially when the fuel is imported. Transportation of fuel implicitly lowers its heating value, an important consideration; the use of local fuel has another added advantage in that it eliminates any fuel security risks.

Most of the devices in common use have the ability to convert energy from one form to another. Table 1.1 shows the 25 possibilities of energy transfer (the symbol x indicates an unrealised link at the present time).

Table 1.1 Energy conversion matrix.

From\To	Mechanical	Electrical	Thermal	Chemical	Nuclear
Mechanical	Gear Nutcracker Push mower	Electric generator	Friction	x	x
Electrical	Electric motor	Light bulb	Electric fire	Electrolysis	Particle accelerator
Thermal	Steam turbine	Thermo-couple	Heat exchanger	x	Fusion reactor
Chemical	Jet engine Rocket	Battery Fuel cell	Car engine Boiler	Intermediate reaction	x
Nuclear	x	x	Nuclear reactor	x	x

1.4 The burning question

Although man discovered fire a very long time ago, the fuel for cooking, heating and light relied on burning just wood. By the 4th century AD, the Romans sent an entire fleet of ships to bring wood from France and North Africa.

It was not until the 17th century that coal was discovered, and this was used extensively during and after the Industrial Revolution in Europe.

In the 19th century, oil was discovered and this has been used extensively by the industrialised world ever since. There have been times when the West has enjoyed cheap oil prices due to competition between oil-producing countries; however, a major shock was felt as a result of the 1973 war between Arab nations and Israel and the subsequent tripling of oil prices.

For all combustion processes, the fuel must contain one or more of the elements (carbon, hydrogen and sulphur). When they react with oxygen, these produce thermal energy – the amount is called its *calorific value*.

Combustion of carbon:

$$C + O_2 \longrightarrow CO_2 + \text{energy}$$

Carbon has an energy content of 32 793 kJ/kg

Combustion of hydrogen:

$$H_2 + 2O_2 \longrightarrow 2H_2O + \text{energy}$$

Hydrogen has an energy content of 142 920 kJ/kg

Combustion of sulphur:

$$S + O_2 \longrightarrow SO_2 + \text{energy}$$

Sulphur has an energy content of 9300 kJ/kg

Table 1.2 Enthalpy 'heat' of combustion of elementary fuel matter.

Substance	Formula	Molecular weight kg/kmol	Heat of combustion kJ/kmol	kJ/kg
Carbon	C (s)	12.001	393 520	32 793
Hydrogen	H_2 (g)	2.016	285 840	142 920
Sulphur	S (s)	32	297 600	9 300
Water vapour	H_2O (g)	18.016	40 620	2256.7

Table 1.2 shows the heat of combustion of these three prime elements for fuels, as they form all combustible products known to mankind. Although the main fuels are coal, oil and natural gas, there are other sources which can be substituted in some cases, such as dry wood, agricultural waste and general household waste.

In the next three subsections, three typical fuels will be analysed, and a method of calculating their respective calorific values and the associated combustion products will be shown.

1.4.1 Combustion of coal

Typical coal in the UK has the following composition (percentage by weight):

Carbon	64.4
Hydrogen	4.4
Oxygen	4.4
Nitrogen	0.9
Sulphur	10
Ash	15
Total	100%

The heat released from burning one kilogram of coal can thus be estimated as the sum of the energy released (Table 1.2) from the proportions given above of carbon, hydrogen and sulphur contained in coal. Hence, the calorific value of coal is calculated as follows:

Part due to combustion of carbon	$+32\,793 \times 0.644 = 21\,119$
Part due to combustion of hydrogen	$+142\,920 \times 0.044 = 6288$
Part due to combustion of sulphur	$+9300 \times 0.1 = 930$
Part lost in combustion of water content	$-2256.7 \times 0.044 - 9 = -893.6$
Calorific value of coal	$21\,119 + 6288 + 930 - 893.6 = 27\,444\,kJ/kg$

This value is obtained on the assumption of 100% efficient and complete combustion.

Emissions:
Production of CO_2 from coal $= 0.644 \times (44/12) = 2.36\,kg\,CO_2/kg$ coal
Production of CO_2 per kWh $= 2.36 \times 3600/27\,444 = 0.309\,kg/kWh$

Production of CO_2 per GJ energy $= 2.36 \times 10^6/27\ 444 = 86.043\ \text{kg/GJ}$
Production of SO_2 from coal $= 0.009 \times (64/32) = 0.018\ \text{kg}\ SO_2/\text{kg coal}$

1.4.2 Combustion of oil

Typical oil in the UK has the following composition (percentage by weight):

Carbon	86
Hydrogen	12
Sulphur	2
Total	100%

The heat released from burning one kilogram of oil can thus be estimated as the sum of the energy released (Table 1.2) from the proportions given above of carbon, hydrogen and sulphur contained in oil. Hence, the calorific value of oil is calculated as follows:

Part due to combustion of carbon	$+32\ 793 \times 0.86$
Part due to combustion of hydrogen	$+142\ 920 \times 0.12$
Part due to combustion of sulphur	$+9300 \times 0.02$
Part lost in combustion of water content	$-2256.7 \times (0.12 \times 9)$
Calorific value of oil	$= 28\ 202 + 17\ 150 + 186$
	$-\ 2437$
	$= 43\ 101\ \text{kJ/kg}$

Emissions:
Production of CO_2 from oil $= 0.86 \times (44/12) = 3.153\ \text{kg}\ CO_2/\text{kg oil}$
Production of CO_2 per kWh $= 3.153 \times 3600/43\ 101 = 0.263\ \text{kg/kWh}$
Production of CO_2 per GJ energy $= 3.153 \times 10^6/43\ 101 = 73.161\ \text{kg/GJ}$
Production of SO_2 from oil $= 0.02 \times (64/32) = 0.040\ \text{kg}\ SO_2/\text{kg oil}$

1.4.3 Combustion of natural gas

Typical natural gas in the UK has the following composition (percentage by weight):

Carbon	75
Hydrogen	25
Sulphur	0
Total	100%

The heat released from burning one kilogram of natural gas can thus be estimated as the sum of the energy released (Table 1.2) from the proportions given above of carbon and hydrogen contained in natural gas. Hence, the calorific value of natural gas is calculated as follows:

Part due to combustion of carbon	$+32\,793 \times 0.75$
Part due to combustion of hydrogen	$+142\,920 \times 0.25$
Part lost in combustion of water content	$-2256.7 \times (0.25 \times 9)$
Calorific value of natural gas	$= 24\,595 + 35\,730 - 5077$
	$= 55\,248\,\text{kJ/kg}$

Emissions:
Production of CO_2 from natural gas $= 0.75 \times (44/12) = 2.75\,\text{kg}\,CO_2/\text{kg gas}$
Production of CO_2 per kWh $= 2.75 \times 3600/55\,248 = 0.179\,\text{kg/kWh}$
Production of CO_2 per GJ energy $= 2.75 \times 10^6/55\,248 = 49.776\,\text{kg/GJ}$
Production of SO_2 from natural gas $= \text{ZERO}$

1.5 Environmental impact from fossil fuels

Coal, oil and natural gas have their relative merits in terms of availability, price and thermal performance. Table 1.3 is based on the calculations carried out in the previous sections and presents a comparison of the heat capacity and CO_2 and SO_2 production of the three fossil fuels. The fourth column is of particular importance in comparing all three fuels; it represents the quantity of carbon dioxide emitted for every unit of energy produced.

Coal produces the highest amount of carbon dioxide for a given output of energy, then oil, then natural gas, which produces nearly half the emissions of coal and a third less than oil.

The results displayed in Table 1.3 for the production of CO_2 mass per unit energy compare well with data published by the UK government. The values found in this chapter are slightly lower than those quoted in the Action on Energy publication; the reason for this is that the calculations shown in this chapter are only concerned with the combustion process itself; these calculations do not take into account the effect of the lifecycle of the fuel, such as the energy used to transport and process the fuel, nor do they include distribution losses.

Table 1.3 Environmental impacts of fossil fuels.

Fuel	Calorific value MJ/kg	CO_2 kg/kg fuel	CO_2/Energy kg/MJ	CO_2/Energy kg/MJ*	SO_2 kg/kg fuel
Coal	26	2.361	0.091	0.093	0.018
Oil	42	3.153	0.075	0.079	0.040
Natural gas	55	2.750	0.050	0.055	0

*Source: Action on Energy Publication No. EEB006.

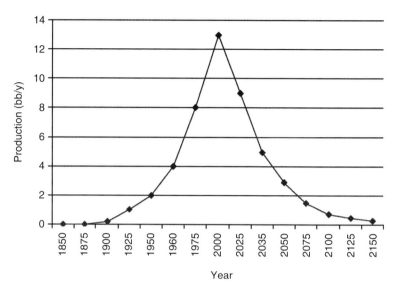

Figure 1.2 The first wake-up call by Hubbert.

1.6 Energy worldwide

The consumption of energy by humankind has evolved over the ages. It began with the invention of fire; man relied on wood burning to cook and to provide warmth and light for millions of years. As civilisation evolved, the needs for energy became greater and other sources were sought. In the long search, man discovered coal. Over the years, coal provided a much greater resource for energy, encouraging man to push its use into further applications.

A major leap in the 19th century was achieved by the discovery of oil in the Middle East. This unfortunate discovery eventually led to two major international wars as the leading industrial nations attempted to dominate the world market and to secure the energy supply for their huge manufacturing industries.

The oil crisis due to the Arab–Israeli war in 1973 resulted in the tripling of oil prices. This was a major shock for nonproducing countries, particularly in Europe, as it put tremendous strain on the energy consumer's budget.

The discovery of oil pushed the competition for manufacturing beyond the industrialised countries' own borders. This competition for shares in the export market put enormous pressure on the reserves of fossil fuel. Hubbert put out a caution when he published his famous curve in 1956 (Figure 1.2). It is clear that the oil reserves of the world are being consumed unsustainably and that they will be exhausted within this century. Humans have to find a new source or sources of energy to replace oil.

However, on the positive side, the depletion of oil can be considered a major advantage to humankind and the environment, in that it will force consumers to reduce their excessive consumption of energy, it will help man to review manufacturing processes and attempt to increase energy efficiency, and it has

Table 1.4   World's energy consumption: past, present and future.

Year	Coal EJ	Oil EJ	Natural gas EJ	Renewable energy EJ	TOTAL EJ	Population millions	Energy kW/capita
1860	3.8	0	0	0	3.8	1000	3.8
1900	20.8	0.8	0.3	0	21.9	1700	12.8
1920	35.8	3.8	0.9	0	40.5	1900	21.3
1940	42.1	11.2	3.1	0	56.4	2000	28.2
1960	60.0	40.2	17.9	10	128.1	2400	53.4
1972	66	115	46	26	253	2500	101.2
1985	115	216	77	33	441	3884	113.5
2000	170	195	143	56	564	5780	97.6
2020	259	106	125	100	590	8846	66.7

probably already pushed governments to search for newer sources of energy. Substantial funds are being allocated to research into renewable resources such as hydropower, wind turbines and solar energy.

Energy consumption worldwide has continued to rise. It is estimated that in 1900, the world consumption was around 22 EJ and by 1960 it had risen to 128 EJ; this reached 564 EJ in 2000 (Table 1.4).

The continued increase in the global population and the associated increase in manufacturing to cater for our greater dependence on energy-driven devices and the culture of multi-car ownership have put even greater importance on energy. It is interesting to note that the energy consumption for individuals increased by tenfold over the course of the 20th century. This is further proof that our use is becoming excessive and that we are much more dependent on energy than we have ever been before.

1.7   Energy and the future

The planet's fossil fuel resource is finite; the production of oil since 1900 is estimated to have reached its peak, and the expected drop in oil reserves will commence shortly. The current rate of consumption is unsustainable; it has taken the Earth millions of years to form these fossil fuels, and it seems that we are about to consume them all in a matter of 200–300 years (see Figure 1.3 and Table 1.5).

The issue is not so much one of 'running out', but more a case of not having enough to keep our current exuberant wasting of energy running. It is a matter of responsibility and obligation to keep some reserves for generations to come.

The second critical issue is that as energy becomes scarce, there will be wars over it. The world has already witnessed major wars in the Middle East. The situation will deteriorate as the fuel supply is reduced.

The shift from fossil fuels to alternative sources is not going to be easy, it is going to take time and much effort and sacrifice in order to adjust.

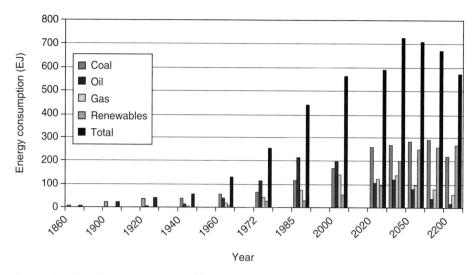

Figure 1.3 World's energy consumption, past and forecast.

Table 1.5 Estimated energy reserves of the world.

Source	Quantity	Form
Earth's daily receipt of solar energy	14.9×10^{21} J	Thermal/ Electromagnetic
Hydropower	3×10^{12} J/s	Mechanical
Tidal power	6.7×10^{10} J/s	Mechanical
Geothermal	0.4×10^{21} J	Thermal
Natural gas	11×10^{21} J	Chemical
Petroleum	11.7×10^{21} J	Chemical
Coal	200×10^{21} J	Chemical
Uranium	1800×10^{21} J	Nuclear fission
Deuterium	$6\,000\,000 \times 10^{21}$ J	Nuclear fusion

As affordable oil is necessary to power any serious attempt at a switchover to alternative sources of energy, extreme prices will severely hinder – if not completely cripple – the ability of society to cope. The economic fallout from high prices will almost certainly produce geopolitical tensions (i.e. war), thereby further hampering the development of large-scale alternative sources of energy. Worse still, in a global environment characterised by massive energy wars, the bulk of the world's financial capital is likely to be disproportionately invested in weapons technologies over alternative energy technologies.

In 1905 Albert Einstein put forward his equivalence theory, and in 1938 Otto Hahn discovered nuclear fission; however, it was not until 1957 that the first nuclear electricity generation was achieved in the USA. Today, France produces 80% of its electricity from nuclear energy.

It is well known by now that although nuclear fission is a valuable source of energy, it has long-term hazards during the production period and the burden of long-term storage of waste after use and the decommissioning of the power plant at the end of its productive life. There has to be a solution in order to avoid such crises; the way forward involves two possibilities.

1.7.1 The dream scenario

A fusion reaction is the process by which large quantities of heat are generated by fusing the nuclei of hydrogen or helium isotopes, which may be derived from seawater. The temperatures and pressures needed to sustain fusion make it a very difficult process to control. In theory, this process has the potential to supply vast quantities of energy with relatively little pollution. Many countries are supporting fusion research; however, inadequate research on quantity and quality has meant no real progress thus far.

If, by extreme good fortune, someone is able to realise nuclear fusion on a realistic scale, it is estimated that this form of energy will sustain life on Earth for millions of years to come. However, realising this goal may be similar, at the moment at least, to someone planning to go to heaven!

1.7.2 The renewable scenario

Renewable energy sources include solar, wind, hydro, geothermal and biomass. All have been realised to a limited degree so far.

This is a more realistic solution for the near future. Renewable energy is freely available and it is pollution-free, but at the present time it does not compete on economic terms with cheap fuels such as coal, oil and gas. This is because, so far, the technology of harnessing renewable energy is relatively inefficient. However, considerable international effort has been coordinated to increase the use of renewable energy in order to reduce global warming caused by the combustion of fossil fuels.

1.8 Worked examples

Worked example 1.1

Determine the 'free' energy of water contained in a lake if the lake size is 2000 m³ and its water level is 300 m above the axis of a water turbine.

Determine the velocity of the water in the supply pipe at the turbine inlet if losses due to friction in the pipe represent the equivalent of 50 m energy head.

Solution:

The potential energy of the water behind the dam is calculated using Equation [1.1]:

Potential energy $= m\,g\,h$

Mass of water $=$ volume of the lake $\times$ density of water

$$= 2000 \times 1000 = 2 \times 10^6\,kg$$

Potential energy, $PE = 2 \times 10^6 \times 9.81 \times 300$

$$= 5.886 \times 10^9\,J$$

Potential energy is converted into kinetic energy at the turbine inlet, less the energy consumed in overcoming friction, hence:

Potential energy $(m\,g\,h) =$ kinetic energy $(\tfrac{1}{2}\,m \times v^2)$

Hence

$$v = \sqrt{2gh} = \sqrt{2} \times 9.81 \times 250$$
$$= 70\,m/s$$

Note that the effective head used is $300 - 50 = 250\,m$ to allow for losses.

Worked example 1.2

A car engine consumes 0.2 litres of fuel every minute; the petrol has a heating value of 42 000k J/kg and a density of 800 kg/m³. If only 30% of the fuel is converted into useful mechanical energy, determine the power of the car. Discuss the fate of the wasted energy.

Solution:

Mass of fuel used $=$ volume $\times$ density

$$= (0.2/1000) \times 800$$
$$= 0.16\,kg/minute$$
$$= 0.0026\,kg/s$$

The chemical energy liberated from burning the fuel is calculated using Equation [1.3]:

Useful energy $=$ mass $\times$ calorific value $\times$ combustion efficiency

$$= 0.0026 \times 42\,000\,000 \times 0.30$$
$$= 33\,600\,J/s\,(W)$$
$$= 33.6\,kW$$

This is the part utilised to move the car. The remaining energy content of fuel has been wasted as:

- thermal energy carried away by the exhaust gases;
- thermal energy carried away by the engine cooling system;
- mechanical energy wasted as friction in various places;
- other minor forms such as sound and light generated by excessive friction between the tyres and the road.

Worked example 1.3

Using Einstein's mass-energy equation ($E = mC^2$)

(a) Determine the energy produced when one milligram of a fissionable material undergoes a nuclear chain reaction.

(b) What is the power rating of such a power plant if the consumption of Uranium is 1kg per day?

Assume the speed of light $C = 3 \times 10^8$ m/s

Solution:

(a) Using Einstein's famous mass-energy equation, the energy liberated from nuclear fission is calculated as follows:

$$E = mC^2$$
$$= 10^{-6} \times (3 \times 10^8)^2$$
$$= 90 \times 10^9 \, J$$

Note that the mass has to be in units of kilograms, and the speed of light in metres per second.

(b) The number of seconds in a day $= 24 \times 3600 = 86\,400$ seconds

$$\text{Power} = E/\text{time} = mC^2/\text{time}$$
$$= 1.0 \times (3 \times 10^8)^2/86\,400$$
$$= 1.04 \times 10^9 \, kW$$

Worked example 1.4

(a) Calculate the thermal energy needed to raise the temperature of 20 litres of water from 20 to 50°C.

(b) If the above process is completed in 4 minutes, what is the power of the heater used?

Assume that, for water, 1 litre weighs 1kg and specific heat capacity $C_p = 4200 \, J/kgK$.

Solution:

(a) Using Equation [1.5] to calculate the heat (thermal energy) required to raise the temperature of water:

$$\text{Thermal energy} = \text{mass} \times \text{specific heat} \times \text{temperature difference}$$
$$= 20 \times 4200 \times (50-20)$$
$$= 2\,520\,000 \, J$$

Note that 20 litres of water has a mass of 20 kg.

(b) By definition, power is the rate of energy consumed, hence:

$$\text{Power} = \text{energy/time}$$
$$= 2\,520\,000/(4 \times 60)$$
$$= 10\,500W \text{ or } 10.5 \, kW$$

Worked example 1.5

Describe the energy conversion in each of the following units:

- a motor car;
- a battery;
- a gas boiler;
- an electric fire.

Solution:

Take, for example, the first part. The motor car converts fuel (chemical energy) into torque (mechanical energy). The other cases are shown in Worked example 1.5, Table 1.

Worked example 1.5, Table 1

Unit	Input	Output
Motor car	Chemical	Mechanical
Battery	Chemical	Electrical
Boiler	Chemical	Thermal
Electric fire	Electric	Thermal

1.9 Tutorial problems

1.1 Describe the energy conversion in each of the following units

- an aeroplane;
- a push mower;
- a radio.

Ans. (chem/mech; mech/mech; elec/mech)

1.2 Calculate the thermal energy needed to raise the temperature of 40 litres of water from 10 to 40°C. If the process is completed in 5 minutes, what is the power of the heater? Assume, for water, 1 litre weighs 1 kg; Cp = 4200 J/kgK.
Ans. (5040 kJ, 16.8 kW)

1.3 Using Einstein's mass–energy equation ($E = mC^2$) determine the energy produced when a one-metre long bar of Uranium of 25 mm diameter (density 8400 kg/m^3) undergoes a nuclear chain reaction. Assume the speed of light, C = 3×10^8 m/s.
Ans. (3.71×10^{17} J)

1.4 A car engine consumes 0.2 litres of fuel every minute; the petrol has a heating value of 45 MJ/kg and a density of 800 kg/m^3. If only one third of the fuel is converted into useful mechanical energy, determine the power of the car.
Discuss the fate of the wasted energy.
Ans. (40 kW)

1.5 Determine the potential energy of water contained in a lake if the lake size is 200 m^3 and its water level is 30 m above the axis of a water turbine. Determine the velocity of the water at the turbine inlet if losses due to friction in the pipe represent the equivalent of 5m energy head.
Ans. (58.86 MJ, 22 m/s)

1.6 Coal has the following composition by weight:

Carbon	50%
Hydrogen	10%
Sulphur	20%
Incombustibles	20%

Determine the calorific value of this fuel assuming that the enthalpy of evaporation for the following is:

Carbon	32 793 kJ/kg;
Hydrogen	142 920 kJ/kg
Sulphur	9300k J/kg
Water vapour	−2256.7 kJ/kg

Ans. (30.5 MJ/kg)

1.10 Case Study: Future energy for the world

The purpose of this case study is to examine current world energy consumption and investigate various strategies for securing a sustainable energy supply. In Case Study 1.10, Table 1, the world energy consumption for the year 2007 (taken from the BP statistics website) is used as a starting point. The idea behind this exercise is to examine a few possibilities in which some of the more polluting fuels are replaced by cleaner fuels, and to debate the merits of each proposal.

Case Study 1.10, Table 1   World energy consumption data 2007.

Type of fuel	World consumption mtoe*	GJ	CO_2 kg/GJ	Total CO_2 million tons	% CO_2 %
Oil	3939	1.6E+11	79	13 029	41.15
Coal	3194	1.3E+11	93	12 437	39.28
Nat gas	2658	1.1E+11	55	6121	19.33
Biomass	1000	4.2E+10	1	42	0.13
Hydro	179	7.5E+09	1	7	0.02
Nuclear	622	2.6E+10	1	26	0.08
Total	11 592	4.9E+11		31 661	100

*mtoe = million tonne oil equivalent.

Note that for hydro and nuclear plants, the CO_2 figures correspond to construction of the plant and manufacture of components.

Now let's propose some schemes for reducing the total emission, discussing the merits of each proposal.

1.10.1   Scheme 1 – Reduce emissions of CO_2 – Replace coal by natural gas

The relevant information for this scheme is shown in Case Study 1.10, Table 2.

Case Study 1.10, Table 2   Reducing emissions by replacing coal by natural gas.

Type of fuel	World consumption mtoe	GJ	CO_2 kg/GJ	Total CO_2 kg	% CO_2 %
Oil	3939	1.6E+11	79	1.3E+13	49.02
Coal	0	0.0E+00	93	0.0E+00	0
Nat gas	5852	2.5E+11	55	1.3E+13	50.7
Biomass	1000	4.2E+10	1	4.2E+10	0.16
Hydro	179	7.5E+09	1	7.5E+09	0.03
Nuclear	622	2.6E+10	1	2.6E+10	0.1
Total	11 592	4.9E+11		2.7E+13	100

Reduction in CO_2 = 16.05%

Merits of the proposal:

For: it provides a 16% reduction in CO_2

Against: expensive solution; finite resource.

1.10.2 Scheme 2 – Reduce emissions of CO_2 – Replace coal by biomass

The relevant information for this scheme is shown in Case Study 1.10, Table 3.

Case Study 1.10, Table 3 Reducing emissions by replacing coal by biomass.

Type of fuel	World consumption mtoe	GJ	CO_2 kg/GJ	Total CO_2 kg	% CO_2 %
Oil	3939	1.6E+11	79	1.3E+13	67.3
Coal	0	0.0E+00	93	0.0E+00	0
Nat gas	2658	1.1E+11	55	6.1E+12	31.62
Biomass	4194	1.8E+11	1	1.8E+11	0.91
Hydro	179	7.5E+09	1	7.5E+09	0.04
Nuclear	622	2.6E+10	1	2.6E+10	0.13
Total	11 592	4.9E+11		1.9E+13	100

Reduction in CO_2 = 38%

Merits of the proposal:

For: it provides a 38% reduction in CO_2

Against: impractical solution as it needs a huge area of plantation. Furthermore, it will affect the land use for food.

1.10.3 Scheme 3 – Reduce emissions of CO_2 – Replace coal by nuclear

The relevant information for this scheme is shown in Case Study 1.10, Table 4.

Case Study 1.10, Table 4 Reducing emissions by replacing coal by nuclear.

Type of fuel	World consumption mtoe	GJ	CO_2 kg/GJ	Total CO_2 kg	% CO_2 %
Oil	3939	1.6E+11	79	1.3E+13	67.3
Coal		0.0E+00	93	0.0E+00	0
Nat gas	2658	1.1E+11	55	6.1E+12	31.62
Biomass	1000	4.2E+10	1	4.2E+10	0.22
Hydro	179	7.5E+09	1	7.5E+09	0.04
Nuclear	3816	1.6E+11	1	1.6E+11	0.83
Total	11 592	4.9E+11		1.9E+13	100

Reduction in $CO_2 = 38\%$

Similar reduction to biomass, but at what environmental cost?

1.10.4 Other possibilities

There is also the possibility of using a combination of renewable energy sources, such as hydro, wind energy, solar energy and biomass.

Chapter 2

Energy Audits for Buildings

Learning outcomes

• Describe the benefits of energy audits	Knowledge and understanding
• Understand the three methods of energy audit	Analysis
• Describe the benchmarking method of energy audit	Problem solving
• Describe the degree-days concept	Knowledge and understanding
• Describe the energy performance certificate	Analysis
• Use the regression method for energy data analysis	Problem solving
• Carry out energy audit, prepare an energy audit report, make recommendations for energy-saving measures	Analysis and reflections
• Practise further tutorial problems	Problem solving

Energy Audits: A Workbook for Energy Management in Buildings, First Edition.
Tarik Al-Shemmeri.
© 2011 Blackwell Publishing Ltd. Published 2011 by Blackwell Publishing Ltd.

2.1   The need for an energy audit

Energy audits initially became popular in response to the energy crisis of 1973 and later years. Interest in energy audits has recently increased as a result of growing understanding of the human impact upon global warming and climate change. It is estimated that 20% of energy consumption is wasted through inefficient energy management.

Good energy management has financial, operational and environmental benefits. Financial benefits may include:

- Reduced fuel and electricity bills.
- Reduced operation and maintenance costs for boilers and other plant, due to fewer hours of operation.
- Reduced capital expenditure on energy conversion plant, partly due to reduced size requirements and partly to a longer operating life of the plant.

Operational benefits for companies, which may indirectly yield financial benefits, may include:

- Improved comfort levels for staff, leading to increased productivity.
- Better information on energy expenditure, which may aid in management decision-making on the allocation of resources among various revenue-saving options.
- Improved public image.

Environmental benefits may include:

- Reduced consumption of finite resources, particularly fossil fuels.
- Reduced greenhouse gas emissions.
- Reduced emission of pollutants, which may otherwise cause local air pollution and acid rain.

The reduction of energy bills is often the main reason for managing energy use. Reductions can range from, perhaps, 10% for organisations already operating efficiently up to 60% or more at sites where the potential had not previously been realised.

A recent study of over 4000 energy surveys by the UK's Energy Efficiency Office showed that average savings of 21% of each site's energy bill were identified, with an average pay-back period for recommendations of 1.5 years.

There are two standard methods commonly used to assess the energy consumption for buildings: *energy benchmarking*, which uses the Normalised Performance Indicator; and *the regression method*.

2.2 The energy benchmarking method

This typically involves a study of energy invoices for a recent representative year, short meetings with one or two key people dealing with energy in the organisation and a quick inspection of the building. The preliminary audit will give an indication of how energy-efficient the building is at present, and the likely potential for cost-effective energy savings. It will also allow the ranking of areas where the potential for energy savings is greatest.

A full energy audit involves a detailed study of energy use in the building. It will cover the energy supply, conversion, distribution, utilisation and rejection to the environment. It will involve a survey of building services and building fabric, meetings with those involved with energy policy and use and the measurement of major energy flows and the performance of major plant. Opportunities for reducing energy consumption and costs are identified and then evaluated to estimate the energy and cost savings that would result.

A standard method for assessing the energy use in buildings is known as the *Normalised Performance Indicator*, which is an overall criterion for consumption of energy indicating how the building compares with others.

In order to be able to conduct such an audit, the following data are required in advance:

- *The annual energy consumption of the building*. This information is most conveniently obtained from past bills, but take care that the figures collected represent actual energy consumed through a full year and are not 'estimated' by the utility. It may be helpful to look at more than one year's bills, provided that there have been no significant changes to the building or its use in that time. The numbers you require are the energy units consumed, not the money value. Include all fuels: natural gas, bottled gas, oil, solid fuel and electricity.
- *The floor area of the building*. This is the total floor area which is directly or indirectly heated, including corridors, toilets and storage areas.
- *The energy used for space heating*. This figure may be difficult to obtain if one energy source is used for more than one purpose and is not metered separately. If this is the case then refer to the guide values given in Table 2.1.
- *The total number of hours that the building is occupied during the year*. Calculate this by multiplying the number of hours in the week for which the building is occupied by the number of weeks in the year for which the building is open, but exclude any cleaning time.
- *Local weather information and building exposure conditions*.

2.2.1 Benchmarking step by step

The following steps describe how to derive the Normalised Performance Indicator for a building:

Table 2.1 Energy conversion factors.

Fuel type	Billed units	To get kWh multiply by
Natural gas	Therms	29.31
	Cubic feet	0.303
Medium fuel oil	Litres	11.3
Heavy fuel oil	Litres	11.4
Coal*	Tonnes	7600
Anthracite*	Tonnes	9200
Liquid petroleum gas (LPG)	Litres	7
	Tonnes	13900

*The calorific value of solid fuel is subject to local variation. A more accurate figure may be available from your supplier.

Step 1: Convert energy units to kWh

Obtain the energy consumption for each fuel over a one-year period from your quarterly bills. Table 2.1 shows the conversion factors that you should use to convert most fuel types into units of kilowatt-hours (kWh).

Step 2: Find the energy used for space heating

When the energy consumed for heating is not known exactly, it may be better to use the recommended figures listed in Table 2.5.

Step 3: Modify the space-heating energy to account for weather

The concept of degree-days is fairly simple. There is an outside temperature, called the *base temperature*, above which heating is not necessary because the occupants of a building will be warm enough due to lighting, sunshine through windows and their own body heat.

Degree-days indicate both the amount of time and the temperature below this base temperature, taken to be 15.5°C. As an example, if, for one week, the average outside air temperature was 12.5°C, this would represent a severity of $(15.5 - 12.5) \times 7 = 21$ degree-days. For a map showing degree-day data across the UK, see Figure 2.1.

In order to calculate the weather correction factor, the total number of degree-days for a 'standard year' is divided by the number of degree-days for the year in which the energy data are to be considered. The standard year is taken to have characteristics that are typical of the last 20 years' average and has 2462 degree-days in it.

$$\text{Weather correction factor} = \frac{\text{Standard degree-days (2462)}}{\text{Degree-days for energy data year}}$$

At this point, multiply the annual space-heating energy obtained in Step 2 by the weather correction factor.

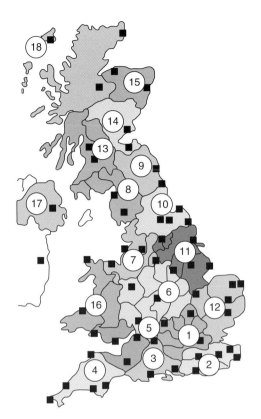

1. Thames Valley
2. South Eastern
3. Southern
4. South Western
5. Severn Valley
6. Midland
7. West Pennines
8. North Western
9. Borders
10. North Eastern
11. East Pennines
12. East Anglia
13. West Scotland
14. East Scotland
15. North East Scotland
16. Wales
17. Northern Ireland
18. North West Scotland

Figure 2.1   UK map showing degree-day regions.

Step 4: Modify the space-heating energy to account for exposure

Part of the heat loss of a building is due to air leaking into and out of the windows and doors. In areas of high exposure it is natural that a building will use more energy to maintain the same internal conditions. Similarly, a well-sheltered building should use less energy. To account for this, an exposure factor is used in a similar manner to the weather correction factor.

From Table 2.2, choose the description of location that most closely describes the site and multiply the space-heating energy by the factor shown.

At the completion of this stage, all the corrections required to the space-heating energy have been made and it is said to be normalised.

Step 5: Add non-heating energy use

All of the other energy use in the building should now be added to the corrected space-heating energy obtained previously. It is not necessary to normalise non-space-heating data, since these are not significantly dependent on weather or exposure. Remember to add all energy used, including separately metered supplies for areas that may have been added to the original building but which were included in your area assessment.

Table 2.2 Exposure factors.

Description of location	Factor
Sheltered The building is in a built-up area with other buildings of similar height or greater surrounding it. This applies to most city-centre locations.	1.1
Normal The building is on level ground in urban and rural surroundings. It is usual to have some trees or adjacent buildings.	1.0
Exposed Coastal and hilly sites with little or no adjacent screening.	0.9

Note: The majority of buildings should be in the normal category.

Step 6: Hours of use factor

To determine this, simply multiply the daily working hours by the number of days the building is occupied or used. Not all buildings are used for the same length of time in a year. To overcome this difficulty, a correction can be made to the annual energy consumption. This works in a manner similar to the degree-day correction and is calculated by reference to Table 2.5. From the table, choose the building type that describes your own facility and note the 'standard hours' alongside. This figure must then be divided by the number of hours actually used, as follows:

$$\text{Hours of use factor} = \frac{\text{Standard number of hours}}{\text{Number of hours for your building}}$$

The hours of use factor reflects the effectiveness of using the building. As with other factors, the result should be fairly close to 1, and should be somewhere between 0.67 and 1.33.

The total energy use obtained in Step 5 should then be multiplied by this factor.

Step 7: Convert the floor area into units of square metres (m²)

Table 2.3 shows the conversion factors to use if the area has been measured in Imperial units.

Table 2.3 Area conversion factors.

To convert from	To m² multiply by
Square feet (ft²)	0.0929
Square yards (yd²)	0.836

Step 8: Calculate the Normalised Performance Indicator (NPI)

In order to calculate the NPI, simply divide the corrected annual energy use, obtained in Step 6, by the area in square metres from Step 7 (the penultimate row in Table 2.4).

$$NPI = \frac{\text{Corrected annual energy consumption}}{\text{Floor area}}$$

This figure describes the amount of energy consumed per unit of floor area within the building under standard conditions.

Step 9: Compare your performance indicator with other buildings

The result of Step 8 is specific to the building under consideration; in order to check how good the building is, it should be compared with other similar buildings (i.e. benchmarked against a number of similar buildings).

Table 2.5 gives yardsticks of performance for buildings. Three bands of performance are shown:

- *Good*. Buildings in this category generally have good controls and energy management procedures, but further energy savings are often still possible.
- *Fair*. Building energy consumption in this band indicates reasonable controls and energy management procedures, but significant energy savings should be achievable.
- *Poor*. Buildings in this band have unnecessarily high consumption and urgent action should be taken to remedy the situation. Substantial energy savings should result from the introduction of energy efficiency measures.

2.2.2   How savings can be achieved

The result of an energy audit will give an indication of the importance of energy management. I have already provided evidence of the necessity to reduce our energy consumption, and even if your result is a fair or good rating, the building will still benefit from further investment to provide even greater reductions in energy wastage.

Savings can be achieved in five main ways:

(1) By altering the physical construction of a building to reduce its heat loss characteristics.
(2) By replacing or upgrading the energy-consuming equipment and controls to make it more efficient.
(3) By changing or modifying energy-consuming equipment to use a less expensive form of energy or more advantageous tariff.
(4) By installing, where a suitable electricity and heat requirement exists, combined heat and power (CHP) plant to meet building electricity and space-heating/hot water needs.

Table 2.4 NPI calculation form.

Convert your energy use into kWh units

Add your quarterly or monthly use over one year for each fuel and enter below

Natural gas	Therms × 29.31 =	kWh
	Cubic ft × 0.303 =	kWh
Medium fuel oil (950sec)	Litres × 11.3 =	kWh
Heavy fuel oil (3500sec)	Litres × 11.4 =	kWh
Coal	Tonnes × 7600 =	kWh
Anthracite	Tonnes × 9200 =	kWh
Liquid petroleum gas (LPG)	Litres × 7 =	kWh
	Tonnes × 13 900 =	kWh
Electricity	kWh × 1 =	kWh
Total energy use for the year	=	kWh A

Find your space-heating energy use

If you can identify any of the fuels above used *only* for space heating, enter the total energy use in kWh

Add these to give total   kWh   B

If you cannot identify these, then choose the space-heating factors applied to the total energy used.

Annual space-heating energy	A × SHF =	kWh C
Annual non-space-heating energy	B or C =	kWh D
	A−D =	kWh E

Adjust the space-heating energy to account for weather

Find the degree-days for the energy data year =   F

The weather correction factor $= \dfrac{2462}{F} = \dfrac{2462}{} =$ G

Adjust the space-heating energy to   kWh   H
standard conditions = D × G =

Adjust the space-heating energy to account for exposure

Obtain the exposure factor from this chapter to suit the location of   J
the building =

Adjusted space-heating energy = H × J   kWh   K

Find normalised annual energy use = E + K =   kWh   L

Correct for hours of use of building

Obtain standard hours of use from this   M
chapter =

Calculate the annual hours of use for your building =   N

Hours of use factor M / N =   P

Annual energy use for standard hours = P × L =   kWh   Q

Find floor area (or pool area) =   m²   R

Find the Normalised Performance Indicator (NPI) = Q/R =   kWh/m²

Compare NPI with yardsticks (Good, Fair or Poor)

Table 2.5 Yardsticks for annual energy consumption of common buildings.

Building type	% Energy used for space heating	Standard hours of use per year	Fair performance range
Nursery	75	2290	370–430
Primary school, no pool	75	1400	180–239
with pool		1480	230–311
Secondary school, no pool	75	1660	189–239
with sports centre		3690	250–280
Special school, non-residential	75	1570	250–339
residential		8760	380–500
University, residential	75	6000	230–314
teaching and admin		3500	189–261
College of further education	75	3200	230–280
Restaurant	60	–	411–520
Public house	60	–	339–470
Air-conditioned offices			
over 2000 m²	60	2600	250–411
under 2000 m²		2400	220–311
Computer centre	60	8760	339–480
Naturally ventilated offices			
over 2000 m²	60	2600	230–289
under 2000 m²		2400	200–250
Swimming pool	55	4000	1050–1389
Sports centre with pool	65	5130	570–840
Sports centre/club, no pool	75	4910	200–340
Library/ Museum/ Art gallery	70	2540	220–312
Church	90	3000	90–170
Small hotels/ guest house	60	–	240–330
Medium-sized hotel	65	–	310–420
Large hotel	70	–	290–420
Bank/ Post office	75	2200	180–240
Cinema	75	3080	650–780
Theatre (public)	75	1150	600–900
Bingo club	75	3500	630–770
Social club	75	3000	200–360
Prison/ Police station/ Fire station	60	8760	550–690
Factory – single shift, 5-day week	80	$8 \times 5 \times 50 = 2000$	180–240
single shift, 7-day week		$8 \times 7 \times 50 = 2800$	230–300
double shift, 5-day week		$16 \times 5 \times 50 = 4000$	260–370
double shift, 7-day week		$16 \times 7 \times 50 = 3920$	277–430

Note: Values below the fair range are considered good, and values above the fair range are considered poor.

(5) By continuous assessment of energy consumption. This can be used to check that both the plant and controls continue to operate as intended, and also to ensure that the occupants' behaviour is not unnecessarily affecting energy use. This is often known as good housekeeping.

Table 2.6 '20-Year' average degree-day data and collection areas.

MONTH	REGION																
	Thames Valley	South Eastern	Southern	South Western	Severn Valley	Midland	North Western	West Pennines	Borders	North Eastern	East Pennines	East Anglia	West Scotland	East Scotland	North East Scotland	Wales	Northern Ireland
Jan	346	368	345	293	321	376	375	361	376	381	372	378	383	388	401	330	365
Feb	322	344	327	285	305	359	345	340	349	358	352	349	352	357	368	320	334
Mar	286	312	301	271	280	322	323	312	330	322	313	317	328	332	346	307	320
Apr	205	233	229	207	201	243	245	230	271	247	232	239	246	263	277	240	242
May	120	150	148	137	128	162	167	144	206	168	154	149	170	197	206	170	171
June	51	74	72	63	56	83	90	75	117	87	78	73	94	109	120	92	92
July	22	39	39	28	24	44	50	38	66	46	42	40	58	62	74	49	53
Aug	25	44	43	28	27	48	56	39	68	49	44	39	64	67	78	45	59
Sept	54	82	79	55	61	90	96	78	104	88	81	71	111	109	127	77	99
Oct	130	160	150	116	138	178	171	157	182	175	165	154	188	192	203	145	173
Nov	242	267	251	206	237	275	284	267	282	281	272	269	299	301	311	235	282
Dec	312	334	312	258	300	343	341	328	339	346	341	341	352	354	362	294	329
Total	2115	2407	2296	1947	2078	2523	2543	2369	2690	2548	2446	2419	2645	2731	2873	2304	2519

2.3   The degree-days concept

Degree-days show how far, for a given month, outside temperatures are, on average, below 15.5°C (the base temperature). When the outside temperature is above this level it should not be necessary to heat a building in normal commercial occupation. This is because the presence of people and office machinery should bring the temperature in the building up to 19°C. The more degree-days there are for a given month, the colder the month. For a 12-month period, a point for each month should be plotted on a graph whose vertical axis shows the amount of fuel used for heating and whose horizontal axis shows degree-days per month. The usual result is a straight line, showing how much heating fuel is required for a given number of degree-days (or a given amount of 'cold'). Any subsequent month's plot which is significantly higher than this '12-month' line may indicate a fault in the heating system. Any point which is lower will confirm an improvement.

Degree-days provide a climatic correction for the prediction of energy consumption over a long time period. The 20-year average degree-day data for various collection areas within the UK are given in Table 2.6. The use of degree-days as a means of compensating for weather conditions has been criticised on the grounds that the method ignores solar radiation, prevailing winds and local outdoor temperature profiles for the building. It may be for these reasons that the correlation coefficient for the best line on the graph of fuel consumption against degree-days is not always high enough.

2.3.1   Regression of degree-day and energy consumption data

Degree-days are used in conjunction with monthly heating and hot water energy data to produce a performance profile. An example is shown in Figure 2.2.

Each point on the graph is the actual monthly energy consumption level correlated with the measured regional degree-day figure.

If the data are correlated for a full year, the straight line function of the energy used for space heating (E) in terms of the degree-days (DD) is given by:

$$E = m\,(DD) + c$$

where: m is the slope of the graph and c is the intercept of the regression equation.

The graph shown in Figure 2.2 was produced by linear regression and is specific to the building under consideration. Utilising this function, the predicted energy consumption for any future time period could be calculated, compared with the actual energy consumed and remedial action taken if necessary.

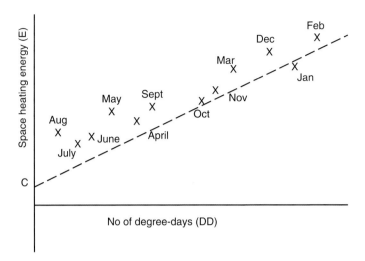

Figure 2.2   Degree-day regression.

Note that if the intercept does not pass through zero, there is a minimum level of energy consumption not connected with heating, e.g. domestic hot water production, catering, etc.

2.4   Energy Performance Certificates

The Energy Performance Certificate (EPC) is the energy efficiency rating of a building as denoted by European Union Directive 2002/91/EC (Figure 2.3). This was inspired by the Kyoto Protocol which aimed to cut back on energy consumption and ecological deterioration due to CO_2 emissions.

In the EPC shown in Figure 2.3, the energy efficiency on the left reflects a measure of the overall use of energy, while the right-hand side shows the associated environmental impact due to the use of energy in that building. For each illustration there are two indicators: one is the actual rating, and next to it is the potential ranking, which can be used as a benchmark for that kind of building. The rating is represented graphically on a scale from A to G, where A stands for the most efficient energy performance and G stands for the least efficient energy performance.

EPCs not only help save money and energy, but also successfully contribute to the cause of environmental improvement. The EPC is a good tool which can be used to help buyers make the right decisions by providing them with adequate information. The better the rating, the higher is the energy efficiency of the property and the lower the fuel bills. With lower energy consumption, expenses will be less and the impact on the climate will be favourable. This being the case, a property with a good energy performance rating will attract more buyers.

Display Energy Certificate

How efficiently is this building being used?

🏛 HM Government

University of Stafford
NELSON LIBRARY
Staffordshire University
Beaconside
STAFFORD
ST18 0AD

Certificate Reference Number:
0361-0610-2909-1526-4006

This certificate indicates how much energy is being used to operate this building. The operational rating is based on meter readings of all the energy actually used in the building. It is compared to a benchmark that represents performance indicative of all buildings of this type. There is more advice on how to interpret this information on the Government's website www.communities.gov.uk/epbd.

Energy Performance Operational Rating

This tells you how efficiently energy has been used in the building. The numbers do not represent actual units of energy consumed; they represent comparative energy efficiency. 100 would be typical for this kind of building.

More energy efficient

76 would be typical

Less energy efficient

A 0-25
B 26-50
C 51-75
D 76-100
E 101-125
F 126-150
G Over 150

Total CO_2 Emissions

This tells you how much carbon dioxide the building emits. It shows tonnes per year of CO_2.

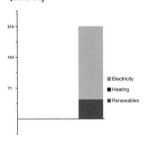

■ Electricity
■ Heating
■ Renewables

06-2010

Previous Operational Ratings

This tells you how efficiently energy has been used in this building over the last three accounting periods

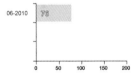

06-2010   76

Technical information

This tells you technical information about how energy is used in this building. Consumption data based on estimates.

Main heating fuel: Natural Gas
Building Environment: Heating and Natural Ventilation
Total useful floor area (m²): 2661
Asset Rating: Not available.

	Heating	Electrical
Annual Energy Use (kWh/m²/year)	87	115
Typical Energy Use (kWh/m²/year)	259	101
Energy from renewables	0%	0%

Administrative information

This is a Display Energy Certificate as defined in SI 2007/991 as amended.

Assessment Software:	i-Prophets Energy Services, digitalenergy, v2.2
Property Reference:	610946520001
Assessor Name:	Richard Hipkiss
Assessor Number:	LCEA095506
Accreditation Scheme:	CIBSE Certification Limited
Employer/Trading Name:	Information Prophets
Employer/Trading Address:	Coventry University Technology Park,Puma Way,Coventry,CV1 2TT
Issue Date:	21-07-2010
Nominated Date:	28-06-2010
Valid Until:	27-06-2011
Related Party Disclosure:	Not related to the occupier

Recommendations for improving the energy efficiency of the building are contained in the accompanying Advisory Report.

Figure 2.3   Energy Performance Certificate.

2.5 Worked examples

Worked example 2.1

The monthly energy data for a bank building is shown in Worked example 2.1, Table 1. Assume that gas is used to provide space heating, while electricity is used for lights and machinery. Also assume that:

- the building is neither exposed nor sheltered;
- hours of use = 9 hours, 5 days, 48 weeks;
- floor area = 50 m × 25 m.

Analyse the performance of the building using the degree-day method. Calculate the NPI for this building and find out if it is energy efficient or not.

Worked example 2.1, Table 1

Month	Natural gas (Therm)	Electricity (kWh)	Degree-days
January	1100	3730	370
February	980	3650	350
March	910	3780	312
April	700	3660	215
May	510	3710	132
June	350	3800	60
July	250	3730	25
August	240	3490	30
September	350	5250	102
October	560	3670	156
November	800	3780	300
December	1020	3410	370

Solution:

First convert all units to kWh, as shown in Worked example 2.1, Table 2.

The regression step has been carried out using a spreadsheet; in this case, Microsoft Excel was used.

Worked example 2.1, Figure 1 represents a linear relation between energy and degree-days. The regression equation, as can be seen from the figure, is:

Gas consumption (kWh) = $67.749 \times DD + 5304.2$

Base load consumption of gas = 5304 kWh.

Worked example 2.1, Table 2

Month	Degree-days (DD)	Natural gas (Therm)	Natural gas (kWh)	Electricity (kWh)
			× 29.31	
January	370	1100	32 241	3730
February	350	980	28 723.8	3650
March	312	910	26 672.1	3780
April	215	700	20 517	3660
May	132	510	14 948.1	3710
June	60	350	10 258.5	3800
July	25	250	7327.5	3730
August	30	240	7034.4	3490
September	102	350	10 258.5	5250
October	156	560	16 413.6	3670
November	300	800	23 448	3780
December	370	1020	29 896.2	3410
Total	2422		227 738.7	45 660

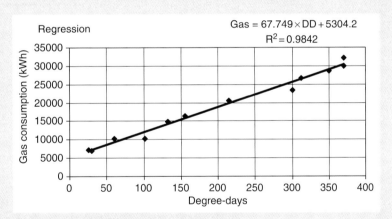

Regression

$$Gas = 67.749 \times DD + 5304.2$$
$$R^2 = 0.9842$$

Worked example 2.1, Figure 1

The NPI method is shown in Worked example 2.1, Table 3.

Worked example 2.1, Table 3

Step 1	Convert all data kWh		
	Gas		227 738.7
	Electricity		45 660
	Total energy	A	273 398.7

(Continued)

Worked example 2.1, Table 3 *(Continued)*

Step 2	SH factor		
		B	227 738.7
	gas heating	C	0
		D	227 738.7
		$E = A - D$	45 660
Step 3	Degree-day correction		
	year/local DD	F	2422
	standard DD	G1	2462
	DD factor	G1	1.02
	adjusted SH	$H = D \times G$	231 499.87
Step 4	Exposure		
	sheltered J =	0.9	
	medium J =	1	
	exposed J =	1.1	
	Input value of J		1
	adjusted SH	$K = H \times J$	231 499.9
Step 5	Normalised energy		
		$L = E + K$	277 159.9
Step 6	Hours standard	M	2200.0
	Hours actual	N	2160.0
	Factor	$P = N/M$	1.019
	Adjusted energy	$Q = P \times L$	282 292.5
Step 7	Area	R	1250.0
Step 8	NPI	Q/R	225.8
Step 9	Yardstick	Range	180 to 239
		Verdict	Fair

Worked example 2.2

The monthly energy data for a sports building with a swimming pool are shown in Worked example 2.2, Table 1. Gas is used to provide space heating as well as to heat the water in the pool, while electricity is used for lights and machinery. Assume that:

- the building is neither exposed nor sheltered;
- hours of use = 12 hours, 7 days, 50 weeks;
- floor area = 30 m × 15 m.

Analyse the performance of the building using the degree-day method. Calculate the NPI for this building and find out if it is energy efficient or not.

Worked example 2.2, Table 1

Month	Natural gas (Therm)	Electricity (kWh)	Degree-days
January	2100	3700	321
February	1980	3600	305
March	1510	3700	280
April	1200	3600	201
May	1110	3700	128
June	1050	3800	56
July	1000	3700	24
August	1020	3500	27
September	1150	5200	61
October	1260	3600	138
November	1500	3700	237
December	2020	3400	300

Solution:

First convert all energy units into kWh, as shown in Worked example 2.2, Table 2 (Therms are multiplied by 29.31 to convert them into kWh).

Worked example 2.2, Table 2

Month	Degree-days (DD)	Natural gas (Therm)	Natural gas (kWh) $\times 29.31$	Electricity (kWh)
January	321	2100	61 551	3700
February	305	1980	58 033.8	3600
March	280	1510	44 258.1	3700
April	201	1200	35 172	3600
May	128	1110	32 534.1	3700
June	56	1050	30 775.5	3800
July	24	1000	29 310	3700
August	27	1020	29 896.2	3500
September	61	1150	33 706.5	5200
October	138	1260	36 930.6	3600
November	237	1500	43 965	3700
December	300	2020	59 206.2	3400
Total	2078		495 339	45 200

The regression equation, as shown in Worked example 2.2, Figure 1 is:

Gas consumption (kWh) = 24 682 + 95.841 $\times$ DD

Base load consumption of gas = 24 682 kWh.

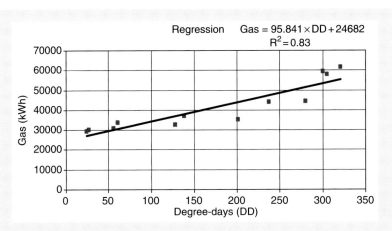

Regression   Gas = 95.841 × DD + 24682
$R^2 = 0.83$

Worked example 2.2, Figure 1

The NPI method is shown in Worked example 2.2, Table 3.

Worked example 2.2, Table 3

Step 1	Convert all data kWh			
	Gas		495 339	
	Electric		45 200	
	Total energy		540 539	A
Step 2	SH factor		0.65	
			0	B
			351 350.35	C
			351 350.35	D
		E = A − D	189 188.65	E
Step 3	Degree-day correction			
	year/local DD		2078	F
	standard DD		2462	G1
	DD factor		1.185	G
	adjusted SH	H = D × G	416 277.46	H
Step 4	Exposure			
	sheltered J =	0.9		
	medium J =	1		
	exposed J =	1.1		
	input value of J		1	J
	adjusted SH	K = H × J	416 277.5	K
Step 5	Normalised energy			
		L = E + K	605 466.1	L
Step 6	Hours standard		5130.0	M
	Hours actual		4200.0	N
	Factor	P = M/N	1.221	P
	Adjusted energy	Q = P × L	739 533.6	Q
Step 7	Area		450.0	R
Step 8	NPI	Q/R	1643.4	
Step 9	Yardstick	Range	570 TO 840	
		Verdict	Poor	

Worked example 2.3

The monthly energy data for a small factory building are shown in Worked example 2.3, Table 1. The following data are given:

- the building is sheltered;
- hours of use = 8 hours (two shifts), 5 days, 50 weeks;
- floor area = 20 m × 20 m.

Analyse the performance of the building using the degree-day method. Calculate the NPI for this building and find out if it is energy efficient or not.

Worked example 2.3, Table 1

Month	Natural gas (kWh)	Electricity (kWh)	Degree-days (DD)
January	3100	3700	321
February	2980	3600	305
March	2510	3700	280
April	2200	3600	201
May	2110	3700	128
June	1500	3800	56
July	1000	3700	24
August	1020	3500	27
September	1150	5200	61
October	2260	3600	138
November	2500	3700	237
December	3020	3400	300

Solution:

First convert all energy units into kWh, as shown in Worked example 2.3, Table 2.

Worked example 2.3, Table 2

Month	Degree-days (DD)	Natural gas (Therm)	Natural gas (kWh)	Electricity (kWh)
			× 29.31	
January	321		3100	3700
February	305		2980	3600
March	280		2510	3700
April	201		2200	3600
May	128		2110	3700
June	56		1500	3800
July	24		1000	3700
August	27		1020	3500
September	61		1150	5200
October	138		2260	3600
November	237		2500	3700
December	300		3020	3400
Total	2078		25 350	45 200

The regression equation, as shown in Worked example 2.3, Figure 1, is:

Gas consumption (kWh) = 981.81 + 6.5295 × DD

Base load consumption of gas = 981.81 kWh.

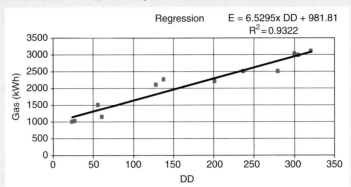

Worked example 2.3, Figure 1

The NPI method is shown in Worked example 2.3, Table 3.

Worked example 2.3, Table 3

Step 1	Convert all data kWh			
	Gas		25 350	
	Electric		45 200	
	Total energy		70 550	A
Step 2	SH factor		0.8	
			0	B
			56 440	C
			56 440	D
		$E = A - D$	14 110	E
Step 3	Degree-day correction			
	year/local DD		2078	F
	standard DD		2462	G1
	DD factor		1.185	G
	adjusted SH	$H = D \times G$	66 869.72	H
Step 4	Exposure			
	sheltered J =		0.9	
	medium J =		1	
	exposed J =		1.1	
	input value of J		1.1	J
	adjusted SH	$K = H \times J$	73 556.7	K
Step 5	Normalised energy			
		$L = E + K$	87 666.7	L
Step 6	Hours standard		4000.0	M
	Hours actual		4000.0	N
	Factor	$P = M/N$	1.0	P
	Adjusted energy	$Q = P \times L$	87 666.7	Q
Step 7	Area		400.0	R
Step 8	NPI	Q/R	219.2	
Step 9	Yardstick	Range	570 TO 840	
		Verdict	Very Good	

2.6 Tutorial problems

2.1 A small factory situated on the outskirts of a town in the Midlands, England has 2523 degree-days in total per year. The factory's working schedule is a single shift of 8 hours per day, 5 days per week for 50 weeks. The energy consumption for the factory is: 1200 MWh natural gas and 50 MWh electricity. The total floor area of the factory is 6000 m². Determine the NPI and assess its energy efficiency.
Ans. (203 kWh/m²)

2.2 The monthly gas consumption figures for a given year for a building are as shown in Tutorial problem 2.2, Table 1, with the degree-days corresponding to the heating season for the region in which the building is sited.

Tutorial problem 2.2, Table 1

Month	Gas consumption (Therm)	Current year (Degree-days)
September	2000	90
October	3500	140
November	8000	250
December	9000	298
January	11 000	400
February	11 500	350
March	10 500	340
April	7500	240
May	6500	180
June	3000	100
July	1000	50
August	1000	44

Plot gas consumption against degree-days and hence estimate the base thermal load.
Ans. (9149 kWh)

2.3 The monthly energy data for a small factory building are shown in Tutorial problem 2.3, Table 1. Assume that the building is sheltered. Hours of use = 8 hours (single shift), 5 days, 50 weeks. The floor area = 30 m × 20 m. Analyse the performance of the building using the degree-day method. Calculate the NPI for this building and find out if it is energy efficient or not.

Tutorial problem 2.3, Table 1

Month	Natural gas (kWh)	Electricity (kWh)	Degree-days
January	3100	3700	401
February	2980	3600	365
March	2510	3700	346
April	2200	3600	277
May	2110	3700	206
June	1500	3800	120
July	1000	3700	74
August	1020	3500	78
September	1150	5200	127
October	2260	3600	203
November	2500	3700	311
December	3020	3400	362

Ans. (111 kWh/m^2, Good)

Chapter 3

Building Fabric's Heat Loss

Learning outcomes

• Describe conduction, convection and radiation heat transfer	Knowledge and understanding
• Distinguish between axial and radial conduction heat transfer	Analysis
• Derive an expression for conduction heat transfer across a typical building's wall construction involving simultaneous conduction and convection	Problem solving
• Solve problems involving conduction, convection and radiation heat transfer	Knowledge and understanding
• Calculate a typical building's heating load	Analysis
• Investigate insulating material, and study current building regulations and insulation standards	Research
• Examine the economics of heating associated with investment in energy efficiency measures and calculate the pay-back period	Analysis and reflections
• Practise further tutorial problems	Problem solving

Energy Audits: A Workbook for Energy Management in Buildings, First Edition.
Tarik Al-Shemmeri.
© 2011 Blackwell Publishing Ltd. Published 2011 by Blackwell Publishing Ltd.

3.1 Modes of heat transfer

Conduction, convection and radiation are the three basic mechanisms by which heat is transferred from a hot source to cooler surroundings. Heat transfer between two media can only take place when the two parties have a finite temperature difference between them.

To give an example of heat transfer, imagine holding one end of a steel rod whose other end is immersed in a fire; before too long, your end of the rod would become hot. There would be a flow of heat along the rod - from the high temperature end to your lower temperature end. This form of heat transfer through a solid material with no apparent movement of the material is called *conduction*.

If you boil an egg in a pan on an electric cooker, the egg is cooked even though it is not in direct contact with the cooker ring. This happens because the water in the bottom of the pan moves upwards as its density decreases with heating and is replaced by colder water coming down from upper layers, and so on. This movement of heated water results in heat being transferred from the hot water to the egg. This mode of heat transfer is called *convection*.

The sun warms the Earth. The heat energy is transferred across a vast distance in space, most of which is a vacuum. This mode of transfer, without involving any intervening medium, is called *radiation*.

Studying the different modes of heat transfer is necessary for two reasons: first, and this may sound contradictory, in order to optimise heat transfer, heat loss needs to be reduced in certain applications such as buildings, while energy transfer is encouraged in other applications such as heat exchangers, e.g. car engine radiators. Secondly, as energy is commonly derived from fuel, saving energy by, for example, insulating buildings, will result in a reduction in energy costs.

3.2 Fourier's law of thermal conduction

Heat transfer by conduction takes place through a solid medium of finite thickness and surface area. Consider a solid material which is heated on one side while the opposite side remains subject to the ambient temperature. The heated side is at a higher temperature than the ambient, and this heat energises the molecules adjacent to the heat source, hence there will be a flow of energy from the hot surface to the cool surface, and this is how thermal energy propagates by conduction.

3.2.1 Conduction through a planar wall

Fourier found that the rate of heat transfer by conduction is proportional to: the area of cross-section, A; the temperature difference, dT and the distance, dx (see Figure 3.1).

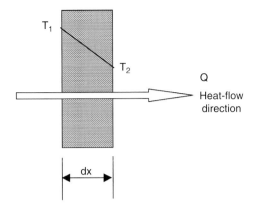

Figure 3.1 Conduction heat transfer.

$$Q \alpha \left(A, \frac{dT}{dx} \right)$$ [3.1]

The constant of proportionality is known as the thermal conductivity of the material through which the heat transfer is taking place

$$Q = -kA \frac{dT}{dx} = -k\,A\,\frac{T_2 - T_1}{x_2 - x_1} = +k\,A\,\frac{T_1 - T_2}{x_2 - x_1}$$ [3.2]

The minus sign appears due to the fact that the temperature is decreasing with increased distance from the hot surface.

3.2.2   Radial conduction through a pipe wall

Consider the flow of hot water through pipes in a central heating network. The radial conduction heat flow through concentric layers of the pipe must be constant, but, since the surface area normal to flow increases with radius, it follows that dT/dx must decrease with radius. The temperature profile is shown in Figure 3.2.

At any radius r, the radial heat conduction is given by:

$$Q = -kA\frac{dT}{dx}$$

where the heat flow direction is along r, and the conduction area is $A = 2\pi rL$

$$Q = \frac{-2\pi\,r\,L\,k\;dT}{dr}$$ [3.3]

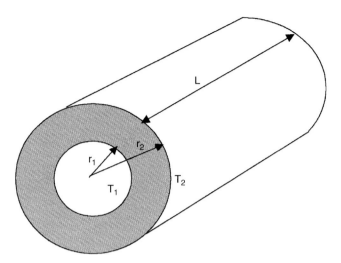

Figure 3.2 Radial conduction.

Integrating between the inner and outer surfaces of the pipe,

$$Q = \frac{-2\pi \; L \, k \; \int_1^2 dT}{\int_1^2 \frac{dr}{r}}$$

The result of the integration yields:

$$Q = \frac{2\pi k \, L \, (T_1 - T_2)}{\ln\left(\frac{r_2}{r_1}\right)}$$

[3.4]

3.3 Heat transfer by convection

Consider the situation where a heated solid surface is kept at a temperature higher than the surrounding air or fluid. This situation is very common, for example it occurs in domestic central heating 'radiators'. In order to analyse such a situation, examine the situation described in Figure 3.3. Imagine the conduction in the solid layer continues into the adjacent fluid layer (gas or liquid); if it can be assumed that the law of heat conduction applies, then

$$Q = -kA\frac{T_s - T_f}{\delta}$$

[3.5]

In the fluid thin film or layer of the ratio of thermal conductivity and the film thickness, k/δ is better known as *the convection heat transfer coefficient*. Hence:

$$Q = h \, A(T_s - T_f)$$

[3.6]

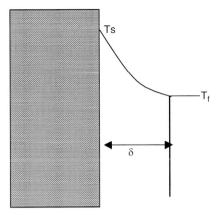

Figure 3.3   Convection heat transfer.

In general, the rate of heat transfer by convection, Q_c, is governed by Newton's equation, Equation [3.6], as the product of the surface area, the temperature difference between the solid surface and the surrounding fluid and a variable known as the convection heat transfer coefficient, h (with units of W/m²K).

There are two types of convection heat transfer: *free/natural convection* and *forced/induced convection*. For example, if a cup of hot coffee is left in a room in the absence of draughts, some of its heat is transferred via density gradients and this represents natural or free convection; whereas if the same cup of hot coffee is left in a room under a ceiling fan, this represents forced or induced convection.

The convection heat transfer coefficient is an empirical quantity determined by experiments and dimensional analysis (correlated by four dimensionless quantities which are defined) and is dependent upon the geometry and properties of the situation. A list of convection relationships is given in the following sections.

The procedure to determine h involves identifying whether it is free convection or forced convection; identifying the geometry; determining whether it is a laminar or turbulent situation and hence identifying an equation for the Nusselt number. Since the Nusselt number (Nu) is a function of the heat transfer coefficient (h), as shown below, h is calculated and the heat transfer by convection can be calculated.

3.3.1   Convective heat transfer: experimental correlations

$$\text{Nusselt No: Nu} = \frac{h\ell}{k}$$

$$\text{Prandtl No: Pr} = \frac{C_p\mu}{k}$$

$$\text{Reynolds No: Re} = \frac{\rho V\ell}{\mu}$$

$$\text{Grashof No: Gr} = \frac{\ell^3 \rho^2 g \, \beta \, \Delta T}{\mu^2}$$

where

h = convective heat transfer coefficient

ℓ = characteristic dimension of a surface (length for a plate or diameter for a tube)

ΔT = surface temp (T_s) – fluid temp (T_f)

g = gravitational acceleration = 9.81 m/s^2

V = mean fluid velocity

Fluid properties are taken at the mean film temperature: $T_m = (T_f + T_s)/2$

k = thermal conductivity

μ = dynamic viscosity

ρ = density

C_p = specific heat at constant pressure

β = $1/T_f$ (coefficient of cubic expression for air; T_f should be in degrees Kelvin).

Material properties can be found from material databases or from technical websites such as Engineeringtoolbox.com.

3.3.2 Free convection

Vertical flat plates and cylinders (ℓ = vertical length of surface):

- Laminar range $10^4 < (\text{Gr, Pr}) < 10^9$, Nu = 0.59 $(\text{Gr, Pr})^{0.25}$
- Turbulent range $10^9 < (\text{Gr, Pr}) < 10^{12}$, Nu = 0.13 $(\text{Gr, Pr})^{0.33}$

Outside horizontal cylinders (ℓ = outside diameter of the cylinder):

- Laminar range $10^4 < (\text{Gr, Pr}) < 10^9$, Nu = 0.53 $(\text{Gr, Pr})^{0.25}$
- Turbulent range $10^9 < (\text{Gr, Pr}) < 10^{12}$, Nu = 0.13 $(\text{Gr, Pr})^{0.33}$

Horizontal flat plate (ℓ = length of longer side), hot plate facing upwards or cold facing downwards:

- Laminar range $10^5 < (\text{Gr, Pr}) < 2 \times 10^7$, Nu = 0.54 $(\text{Gr, Pr})^{0.25}$
- Turbulent range $2 \times 10^7 < (\text{Gr, Pr}) < 3 \times 10^{10}$, Nu = 0.14 $(\text{Gr, Pr})^{0.33}$

3.3.3 Forced convection

Flow along flat plates (ℓ = length of plate in the direction of flow):

- Laminar range   Re < 10^5, Nu = 0.664 $\text{Re}^{0.5}$ $\text{Pr}^{0.33}$. Correlation is suitable for 0.6 < Pr < 10 and heat is transferred over the whole of the plate.

- Turbulent range $Re > 10^5$, $Nu = 0.037 \, Re^{0.8} \, Pr^{0.33}$. Correlation is suitable for $Pr > 0.6$ and heat is transferred over the whole of the plate.

Flow inside tubes (ℓ = inside diameter of the tube):

- Laminar range $Re < 2500$, $Nu = 4.1$

Fully developed flow, heating or cooling:

- Turbulentt range $Re \geq 2500$, $Nu = 0.023 \, Re^{0.8} \, Pr^n$. Suitable for $0.7 < Pr < 100$.

where $n = 0.4$ if fluid is being heated; $n = 0.3$ if fluid is being cooled.

External flow over cylinders (ℓ = cylinder outside diameter):
Across isolated cylinders and tubes:

- $1 < Re < 4000$, $Nu = 0.43 + 0.53 \, Re^{0.50} \, Pr^{0.31}$
- $4000 < Re < 40\,000$, $Nu = 0.19 \, Re^{0.62} \, Pr^{0.31}$
- $40\,000 < Re < 400\,000$, $Nu = 0.027 \, Re^{0.81} \, Pr^{0.30}$

Across banks of cylinders or tubes:

- $2000 < Re < 32\,000$, $Nu = b \, Re^{0.6} \, Pr^{0.33}$

where $b = 0.33$ for staggered tubes and $b = 0.26$ for tubes in line.

3.4 Heat transfer through a composite wall separating two fluids

In buildings where there is a finite difference between the inside and outside air temperatures, the situation in terms of heat transfer constitutes a mixture of conduction through the solid envelope and convection on either side of the walls, roof, etc. Heat transfer between two fluids kept at T_a and T_b and separated by solid layer/s 'x' distance/s apart, consists of convection from fluid a followed by conduction through the three solid layers and finally convection to the outer fluid, b (Figure 3.4).

$$\therefore \frac{Q}{A} = h_a \, (T_a - T_1) \qquad \text{convection from air to face 'a'}$$

$$= \frac{k_1 (T_1 - T_2)}{x_1} \qquad \text{conduction across layer } x_1$$

$$= \frac{k_2 (T_2 - T_3)}{x_2} \qquad \text{conduction across layer } x_2$$

$$= \frac{k_3 (T_3 - T_4)}{x_3} \qquad \text{conduction across layer } x_3$$

$$= h_b \, (T_4 - T_b) \qquad \text{convection from air to face 'b'}$$

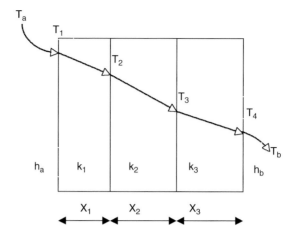

Figure 3.4 Conduction through multiple layers.

Since the heat flux (Q/A) is constant, we can rewrite the five equations in terms of temperatures and add them, to give:

$$T_a - T_b = \frac{Q}{A}\left[\frac{1}{h_a} + \frac{x_1}{k_1} + \frac{x_2}{k_2} + \frac{x_3}{k_3} + \frac{1}{h_b}\right]$$

$$\therefore \frac{Q}{A} = \frac{(T_a - T_b)}{\left(\dfrac{1}{h_a} + \dfrac{x_1}{k_1} + \dfrac{x_2}{k_2} + \dfrac{x_3}{k_3} + \dfrac{1}{h_b}\right)}$$

or

$$Q = AU\,(T_a - T_b) \tag{3.7}$$

where U is the overall heat transfer coefficient:

$$U = \frac{1}{\left[\dfrac{1}{h_a} + \dfrac{x_1}{k_1} + \dfrac{x_2}{k_2} + \dfrac{x_3}{k_3} + \dfrac{1}{h_b}\right]} \tag{3.8}$$

3.5 Heat exchange through a tube with convection on both sides

Consider a pipe carrying a fluid (liquid or gas) which is kept at a temperature higher than the ambient. This situation encompasses radial conduction through the pipe material, as well as convection on either side of the pipe's surface.

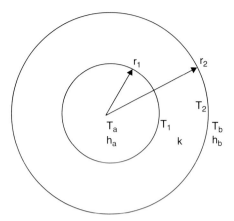

Figure 3.5 Combined convection–conduction–convection heat transfer.

With reference to Figure 3.5, the heat transfer through a pipe separating two fluids (i.e. conduction and convection on the inside and outside layers) is given by:

$$Q = \frac{2\pi k L(T_2 - T_1)}{\ell n(r_2 / r_1)}$$

$$Q = A_1.h_a\left(T_1 - T_a\right)$$

$$Q = A_2.h_b\left(T_b - T_2\right)$$

Rewriting

$$Q = \frac{T_a - T_b}{\dfrac{1}{2\pi r_1 L\, h_a} + \dfrac{\ell n(r_2/r_1)}{2\pi r_1 L\, k} + \dfrac{1}{2\pi r_2 L\, h_b}}$$

The heat transfer can be written in relation to the inside area as follows:

$$Q = \frac{A_i \times (T_a - T_b)}{\dfrac{1}{h_a} + \dfrac{r_1 \times \ell n(r_2/r_1)}{k} + \dfrac{(r_1 / r_2)}{h_b}} \qquad [3.9]$$

3.6   A composite tube with fluid on the inner and outer surfaces

Composite tube geometries are found in situations when a relatively hot or cold fluid passes through a markedly different ambient temperature. In other words, composite tubes are required in situations where insulation is beneficial in order to save energy.

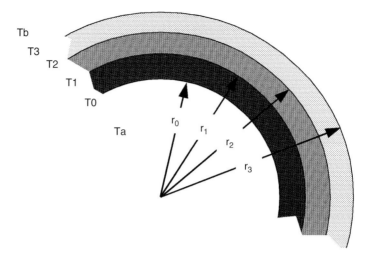

Figure 3.6 Radial conduction through multiple layers.

Consider Figure 3.6. Heat transfer consists of four components: convection on either side of the solid surfaces, and conduction through the pipe material and the insulation material. Hence:

$$Q = 2 \pi L r_1 h_a (T_a - T_0) \qquad \text{inside film}$$
$$= 2 \pi L k_1 (T_0 - T_1)/ \ln(r_1/r_0) \qquad \text{first layer conduction}$$
$$= 2 \pi L k_1 (T_1 - T_2)/ \ln(r_2/r_1) \qquad \text{second layer conduction}$$
$$= 2 \pi L k_2 (T_2 - T_3)/ \ln(r_3/r_2) \qquad \text{third layer conduction}$$
$$= 2 \pi L r_3 h_b (T_3 - T_b) \qquad \text{outside film}$$

Rearranging the above equations gives:

$$Q = \cfrac{T_a - T_b}{\cfrac{1}{2\pi r_0 L h_a} + \cfrac{\ln(r_1/r_0)}{2\pi k_1 L} + \cfrac{\ln(r_2/r_1)}{2\pi k_2 L} + \cfrac{\ln(r_3/r_2)}{2\pi k_3 L} + \cfrac{1}{2\pi r_3 L h_b}} \qquad [3.10]$$

3.7 Heat transfer by radiation

All bodies radiate thermal energy. If two bodies have different temperatures and the space between them is unoccupied, the hotter body will emit radiation and the colder body will absorb part of this radiant energy. Thermal radiation, like light, travels at the same speed through a vacuum, and this is the secret of our heat energy received from the sun.

The net outcome of heat exchange by radiation between two bodies is given by the Stefan-Boltzmann equation:

$$Q_r = e \, \sigma \, F_{12} \, A_1 \, (T_1^4 - T_2^4) \qquad [3.11]$$

where T_1 and T_2 are, respectively, the absolute temperatures (K) for the hot surface and ambient air; A_1 is the surface area of the emitter; $\sigma = 5.67 \times 10^{-8}$ W/m²K⁴; F_{12} is the shape factor (for small surfaces emitting heat in a large space, $F_{12} = 1$); and e is the emissivity for a hot surface radiating to atmosphere. If two surfaces (with emissivity values e_1 and e_2) are exchanging radiant heat, then effective emissivity is calculated as:

$$ e = \frac{1}{\dfrac{1}{e_1} + \dfrac{1}{e_2} - 1} \qquad [3.12] $$

Thermal energy falling onto a surface may be absorbed, reflected and/or transmitted depending on the nature of the surface. Some gases transmit nearly all the radiant heat. A body that absorbs all the impinging radiant heat is called a *black body* – a hypothetical conception in which the absorptivity is unity. The nearest approach to a black body is obtained by a hollow vessel penetrated only by a small pin hole through which radiant heat may pass to the inside; once inside, little of this radiation is reflected back through the pin hole.

The label 'black body' has nothing to do with the colour, and a painted white surface may absorb about the same total radiation as a painted black surface. However, surfaces have a certain selectivity regarding reflection; over visible wavelengths, a white painted surface reflects a large part whereas a black surface absorbs a large part of the incident energy, and so it is considered a good heat sink.

Since absorptivity is a surface property, accumulation of dust, corrosion, etc. may have drastic effects on heat transfer rates by radiation. Brightly polished metals are such good reflectors that most of the radiant heat may be reflected.

3.8 Building fabric's heat load calculations

How do we calculate heat loss or heat gain due to a temperature difference across a building element (this includes all walls, the roof, floor, door/s and windows)?

The general equation for heat loss/gain through a building's structure (wall or roof) is given in terms of the side area (A), its U value and the temperature difference between the inside and outside air temperatures, as follows:

$$ Q_F = A \, U \, \Delta T \qquad [3.13] $$

The overall conductance (U) is usually tabulated for common building materials (see Table 3.1).

However, the U value can be determined for nonstandard structures, by adding the contribution of all elements of the structure. Note that every building

Table 3.1 *U* values for common types of construction.

Construction	U value (W/m²K)
Walls (brickwork)	
220 mm solid brick wall unplastered	2.3
220 mm solid brick wall with 16 mm plaster	2.1
260 mm unventilated cavity wall with 105 mm brick outer and inner leaves, with 16 mm plaster	1.5
260 mm unventilated cavity wall with 105 mm brick outer leaf and 100 mm lightweight concrete block inner leaf and 16 mm plaster	0.96
260 mm cavity wall as above but with 13 mm expanded polystyrene board in cavity	0.70
Flat roofs	
19 mm asphalt on 150 mm solid concrete	3.4
19 mm asphalt on 150 mm hollow tiles	2.2
19 mm asphalt on 13 mm cement and sand screed, 50 mm woodwork slabs on timber joists and aluminium foil backed by 10 mm plasterboard ceiling	0.9
Pitched roofs	
Tiles on battens, sarking felt and plasterboard ceiling	2.44
Tiles on battens, sarking felt and aluminium foil-backed 10 mm plasterboard ceiling	1.50
As above but with 100 mm glassfibre insulation laid between joists	0.35
Floors	
Solid ground floor with exposed edges 3 m by 3 m	1.47
Solid ground floor with exposed edges 30 m by 30 m	0.26
Solid ground floor with exposed edges 60 m by 60 m	0.11
Intermediate floor on joists with plaster ceiling	1.65
Intermediate floor with 150 mm concrete, 50 mm screed and timber flooring	1.87
Suspended timber floor above ground 3 m by 3 m	1.05
Suspended timber floor above ground 30 m by 30 m	0.28
Suspended timber floor above ground 60 m by 60 m	0.16
Glass	
Single glazing	6.0
Double glazing with 6 mm air space (13 mm)	3.5 (3.0)
Triple glazing with 6 mm air space (13 mm)	2.5 (2.0)

wall is located between two air layers – the inside and outside air – and each has an effect on the flow of heat.

$$U = \frac{1}{R_{in} + \Sigma(x/k) + R_{air} + R_{out}}$$ [3.14]

where R_{in} and R_{out} are the thermal resistance of the air on the inside and outside surfaces respectively, and R_{air} is the air cavity resistance if not filled. The summation symbol represents the addition of thermal resistances of all layers of materials making up the wall. Typical values for air thermal resistance for walls are $0.12\,m^2\,K/W$ on the inside and $0.06\,m^2\,K/W$ on the outside; for an air cavity the value is $0.10\,m^2\,K/W$.

3.9 Energy efficiency and the environment

As the cost of energy continues to rise, the need to examine the various ways in which savings can be made becomes ever more important. Reducing energy consumption has two benefits: from an economic point of view, it reduces the costs involved, and from an environmental point of view, it reduces emissions and helps towards sustainability.

3.9.1 Space heating

Space heating is concerned with raising the temperature of the air in the working space above that of the outside ambient temperature. Space heating can be achieved either by direct means, such as electrical heaters and open fires (coal, oil or gas), or indirectly via a central heating boiler (again this could use solid, liquid or gas fuels) heating water to heat the air in the working space by use of radiators. The heat required to raise the temperature of air is calculated by:

$$Q = \text{Mass of air} \times \text{Specific heat for air} \times \text{Temperature difference} \qquad [3.15]$$

This heating requirement is to compensate for the heat lost through the structure of the building and ventilation. Solar heat gain through windows provides a valuable source of heat contribution to space heating.

Heat lost through windows is given by:

$$Q_w = \text{Area of window} \times U \text{ value} \times \text{Temperature difference} \qquad [3.16]$$

An investment in good quality windows may pay for itself in time. Generally, a double glazed window has nearly half the U value of a single glazed window, and unless the cost is prohibitively high, such investment is highly recommended.

The heat lost through a building's surface can be expressed simply as:

$$\text{Fabric heat loss/m}^2 = U \text{ value} \times \text{Temperature difference}$$

The temperature difference for an annual heating season in the UK, for example, is estimated at 2259 degree-days, where, to recap, a degree-day is the number of days multiplied by the difference in temperature if it falls below 15°C. Hence, when multiplied by 24 hours, the fabric heat loss can be expressed in units of $kWh/m^2/year$ as:

$$Q_F = 54.216 \times U \qquad [3.17]$$

The annual heat loss due to ventilation for a typical house is usually approximated as:

$$Q_v = 36\,kWh/m^2 \qquad\qquad [3.18]$$

U values are usually reported for a typical building with standard average exposure. Multiply by an exposure coefficient of 0.9 for sheltered buildings, and 1.1 for exposed buildings.

3.9.2 Insulation standards

It is evident that, in our drive for energy efficiency and in order to reduce the cost of heating, there is an incentive to improve the insulation of a building; Table 3.2, for example, shows the heating load for a typical building without insulation.

Until recently, the standards for insulation were fairly relaxed, but international pressure and the commitment to reducing CO_2 emissions have led to the introduction of new standards for insulation. Table 3.3 shows these standards for U values for typical buildings.

3.9.3 The economics of heating

Table 3.3 demonstrates clearly the huge reduction (and consequent saving) in waste heating energy from buildings. Simply by changing the structure of the

Table 3.2   Heating load for a typical building with no insulation.

Item	U value (W/m²K)	Heating (KWh/ m²/year)	Percent
Lofts	1.5	81.3	38.17
Walls	0.96	52.0	24.41
Floors	1.47	79.70	37.42
Total =		213kWh/m²	100%

Table 3.3   Building Regulations – U values (W/m²K).

Item	Uninsulated	UK standards, pre 2006	UK standards, post 2006
Lofts	1.5	0.35	0.12
Walls	0.96	0.6	0.25
Floors	1.47	0.70	0.20
Windows	6.0	3.0	2.0
	Single glazed	Double glazed	Triple glazed

building's shell, it has been possible to reduce the building heating load from 213 to 31 kWh/m² per year - a huge saving of 85%, i.e. the heating load has been reduced to under 15% of that 20 years ago.

There is so much more that can be done, especially with domestic equipment; whether buying a new item or retrofitting an appliance or lights, it is important to consider three factors:

- the capital cost;
- the operational costs of running the unit (i.e. cost of fuel);
- the cost of maintenance.

The simple pay-back period will be used to demonstrate whether or not investment in an energy-saving scheme is feasible.

$$\text{Pay-back} = \frac{\text{Capital}}{\text{Saving/year}} \qquad [3.19]$$

The pay-back period principle takes essentially an 'investment' view of the action, plan or scenario, and its estimated cash flow stream. The pay-back period is the length of time required to recover the cost of an investment (e.g. purchase of more efficient equipment or lights), usually measured in years. Other things being equal, the better investment is the one with the shorter pay-back period.

Pay-back periods are sometimes used as a way of comparing alternative investments with respect to risk. The investment with the shorter pay-back period is considered less risky. As an example, consider a £250 hot water boiler purchase that is expected to reduce energy consumption by £50 per year:

Pay-back period = £250/£50 year^{-1} = 5 years

From year 6 onwards, there will be a net profit of £50 per year for as long as the boiler keeps working (a typical boiler's life expectancy is 20 years on average).

The pay-back period is an appealing method because it is easily interpreted and understood. Nevertheless, it has some limitations: a pay-back cannot be calculated if the positive cash inflows do not eventually outweigh the cash outflows. This is why pay-back is of little use when used with a pure 'costs only' business case or cost of ownership analysis.

A pay-back calculation ordinarily does not recognise the time value of money (in a discounting sense) nor does it reflect money coming in after pay-back (contrast this with discounted cash flow and internal rate of return).

It is usually assumed that the longer the pay-back period, the more uncertain are the positive returns. For this reason, the pay-back period is often used as a measure of risk, or a risk-related criterion that must be met before funds are spent. However, this does not account for the additional environmental benefits resulting from saving energy.

3.10 Worked examples

Worked example 3.1

A composite wall is made up of external brickwork with a thick layer of fibreglass. The fibreglass is faced internally by a thick insulating board.

Brickwork (110mm) $k = 0.60$ W/mK
Fibreglass (75mm) $k = 0.04$ W/mK
Insulating board (25 mm) $k = 0.06$ W/mK

The surface heat transfer coefficient on the inside wall is 2.0 W/m²K while that on the outside wall is 3.0 W/m²K. Determine the overall heat transfer coefficient for the wall and the heat lost through such a wall 6m high and 10m long. Take the internal ambient temperature to be 20°C and the external ambient temperature to be 10°C.

Solution:

Refer to Worked example 3.1, Figure 1. The overall heat transfer coefficient

$$U = \cfrac{1}{\left(\cfrac{1}{h_i} + \cfrac{x_1}{k_1} + \cfrac{x_2}{k_2} + \cfrac{x_3}{k_3} + \cfrac{1}{h_o} \right)}$$

$$= \cfrac{1}{\left(\cfrac{1}{2.0} + \cfrac{0.025}{0.06} + \cfrac{0.075}{0.04} + \cfrac{0.110}{0.6} + \cfrac{1}{3.0} \right)}$$

$$= \cfrac{1}{0.5 + 0.417 + 1.875 + 0.183 + 0.333}$$

$$= 0.302 \ W/m^2K$$

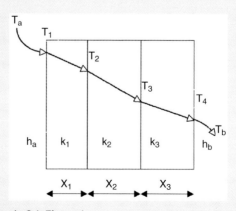

Worked example 3.1, Figure 1

The heat transfer rate

$$Q = U A (T_i - T_o)$$
$$= 0.302 \times 6 \times 10 \times (20 - 10) = 181\,W$$

Worked example 3.2

A cold store is used to keep food products at a temperature of −5°C while the ambient temperature is 15°C. The convective heat transfer coefficients on the inside and outside of the store walls are 10 and 20 W/m²K respectively. The thermal conductivity of the wall is 0.8 W/mK and its thickness is 40 cm.

(a) Calculate the heat transfer rate across the store, assuming its dimensions are $5 \times 3 \times 2\,m$.
(b) The owners wish to save energy by adding an insulating material on the inside; the insulator has a thermal conductivity of 0.05 W/mK. Determine the required thickness of insulation required to provide a 20% reduction in the heat transfer rate.

Solution:

(a) The overall h.t.c.

$$U = \frac{1}{\dfrac{1}{h_i} + \dfrac{x}{k} + \dfrac{1}{h_o}} = \frac{1}{\dfrac{1}{10} + \dfrac{0.4}{0.8} + \dfrac{1}{20}} = 1.538\ W/m^2K$$

Conduction area of the store $A = 2 \times [5 \times 3 + 5 \times 2 + 3 \times 2] = 62\,m^2$
The heat transfer rate is:

$$Q = A U (T_i - T_o)$$
$$= 62 \times 1.538 \times (15 - (-5))$$
$$= 1908\,W$$

(b) To save 20% on heating, the new heat transfer rate $= 0.8 \times Q1 = 1526\,W$

$$Q_{new} = \frac{A.\,dT}{\dfrac{1}{h_i} + \left[\dfrac{x}{k}\right]_{wall} + \left[\dfrac{x}{k}\right]_{ins} + \dfrac{1}{h_c}}$$

$$1526 = \frac{62 \times 20}{\dfrac{1}{10} + \dfrac{0.4}{0.8} + \dfrac{x}{0.05} + \dfrac{1}{20}}$$

$$x = 8\,mm$$

Worked example 3.3

A 3mm-thick steel pipe with internal diameter 24 mm carries steam at 320°C. The thermal conductivity of the pipe material is 50. The inside and outside convective heat transfer coefficients are 100 and 10 W/m²K, respectively. The outside air temperature is 20°C. Calculate:

(a) The heat transfer per metre length from the uninsulated pipe.
(b) The reduction in thermal energy losses per metre length of the pipe achieved by a 10 mm-thick insulation material ($k = 0.03$ W/mK) wrapped on the outside of the pipe.

Solution:

(a) The uninsulated pipe:

$$U_i = \frac{1}{\dfrac{1}{h_i} + \dfrac{r_i \times \ln(r_2/r_1)}{k_1} + \dfrac{1}{h_o} \cdot \dfrac{r_i}{r_o}}$$

$$= \frac{1}{\dfrac{1}{100} + \dfrac{0.012 \times \ln(0.015/0.012)}{50} + \dfrac{0.012}{10 \times 0.015}} = 11\,\text{W/m}^2\text{K}$$

$$Q = A.U.dT = (2\pi \times 0.015 \times 1) \times 11 \times (320{-}20) = 311\,\text{W/m}$$

(b) The insulated pipe:

$$U_i = \frac{1}{\dfrac{1}{h_i} + \dfrac{r_1 \times \ln(r_2/r_1)}{k_1} + \dfrac{r_1 \times \ln(r_3/r_2)}{k_2} + \dfrac{1}{h_o} \cdot \dfrac{r_1}{r_o}}$$

$$= \frac{1}{\dfrac{1}{100} + \dfrac{0.012 \times \ln(0.015/0.012)}{50} + \dfrac{0.012 \times \ln(0.025/0.015)}{0.03} + \dfrac{0.012}{10 \times 0.025}}$$

$$= 3.8\,\text{W/m}^2\text{K}$$

$$Q = A.U.\Delta T = (2\pi \times 0.015 \times 1) \times 3.8 \times (320{-}20) = 107\,\text{W/m}$$

Quite a big difference – only one-third of the original heat loss, or a saving of 65%.

Worked example 3.4

A central heating panel (with emissivity of 0.85) is placed by a brick wall (with emissivity of 0.93). The surface temperature of the heating panel is kept at 80°C while the brick surface is 20°C.

(a) Determine the radiation heat exchange per unit area between the two surfaces.
(b) In order to reduce the radiation between the two surfaces, a sheet of aluminium is placed in parallel between the two surfaces. If aluminium has an emissivity of 0.05, determine the percentage reduction in radiation heat transfer.

Solution:

(a) No foil, with $A_1 = 1.0$; $\sigma = 5.67 \times 10^{-8}\,W/m^2K^4$; $F_{12} = 1$.

$$e = \frac{1}{\dfrac{1}{e_1} + \dfrac{1}{e_2} - 1} = \frac{1}{\dfrac{1}{0.85} + \dfrac{1}{0.93} - 1} = 0.799$$

Hence

$$\begin{aligned} Q_r &= e\,\sigma\,F_{12}\,A_1\,(T_1^4 - T_2^4) \\ &= 0.799 \times 5.67 \times 10^{-8} \times 1 \times 1 \times [(80 + 273)^4 - (20 + 273)^4] \\ &= 369.5\,W/m^2 \end{aligned}$$

(b) When the foil is used, the exchange of radiation is between the central heating panel and the foil. Hence:

$$e = \frac{1}{\dfrac{1}{e_1} + \dfrac{1}{e_{foil}} - 1} = \frac{1}{\dfrac{1}{0.85} + \dfrac{1}{0.05} - 1} = 0.049$$

Hence

$$\begin{aligned} Q_r &= e\,\sigma\,F_{12}\,A_1\,(T_1^4 - T_2^4) \\ &= 0.049 \times 5.67 \times 10^{-8} \times 1 \times 1 \times [(80 + 273)^4 - (20 + 273)^4] \\ &= 22.9\,W/m^2 \end{aligned}$$

Hence, the percentage reduction $= 100 \times (369.5 - 22.9)/369.5 = 94\%$.

Worked example 3.5

An electric heater consisting of a single horizontal bar with diameter 25mm and length 0.3m is used to maintain a room temperature of 17°C. If the heater surface temperature is 537°C, calculate:

(a) The convective heat transfer.
(b) The radiation heat transfer, if the heater surface has an emissivity of 0.6.
(c) The total heat transfer.

Use the following data:

Properties of air:
$C_p = 1.0398\,kJ/kgK$, $\mu = 2.849 \times 10^{-5}\,kg/ms$,
$k = 4.357 \times 10^{-5}\,kW/mK$, $\rho = 0.6418\,kg/m^3$

Solution:

(a) This situation involves free convection, hence calculate Gr and Pr. $(\beta = 1/T_a = 1/(273 + 17) = 3.448 \times 10^{-3})$

$$Gr = \frac{\rho^2 d^3 \beta g \Delta T}{\mu^2} = \frac{0.6418^2 \times 0.025^3 \times 3.448 \times 10^{-3} \times 9.81 \times (537 - 17)}{(2.849 \times 10^{-5})^2}$$
$$= 139\,478$$

$$Pr = \frac{\mu Cp}{k} = \frac{2.849 \times 10^{-5} \times 1039.8}{4.357 \times 10^{-2}} = 0.68$$

check Gr Pr $= 139\,478 \times 0.68 = 94\,845 < 10^9$

$\therefore$ Nu $= 0.53\,(Gr\,Pr)^{0.25} = 0.53\,(94\,845 < 10^9)^{0.25} = 9.3$

Nu $= \dfrac{h\,d}{k}$ $\therefore h = 9.3 \times 4.357 \times 10^{-2}/0.025 = 16.21\,W/m^2K$

$\therefore Q_{conv} = h\,A\,\Delta T = 16.21 \times \pi \times 0.025 \times 0.3\,(537 - 17) = 198.6\,W$

(b) The radiation heat transfer is:

$$Q_{rad} = e\,\sigma\,F_{12}\,A_1\,(T_1^4 - T_2^4)$$
$$= 0.6 \times 5.67 \times 10^{-8} \times 1 \times \pi \times 0.025 \times 0.3\,[(537 + 273)^4 - (17 + 273)^4]$$
$$= 339\,W$$

(c) $\therefore Q_{total} = Q_{conv} + Q_{rad} = 198.6 + 339 = 537.6\,W$

Worked example 3.6

A domestic central heating system circulates water at 80°C to maintain a mean air temperature of 18°C. The system consists of ten radiators of height 1m and length 1m. Assume that the emissivity of the radiators is 0.9. Calculate the mean heat input to the boiler per hour.

Use the following data:

Air properties:
$k = 2.816 \times 10^{-2}$ W/mK
$Pr = 0.701$
$\mu = 1.962 \times 10^{-5}$ kgm^{-1}s^{-1}
$\rho = 1.086$ kgm^{-3}

For free convection from a vertical flat plate, the empirical relation is applicable:

$Nu = 0.13 \times (Pr\ Gr)^{0.33}$

Solution:

The dimensionless groups are calculated at mean temperature:

$$Pr\,Gr = \frac{pr.g\ell^3 \Delta T \beta}{v^2} = \left[\frac{0.701 \times 9.81 \times 1^3 \times (80-18)}{\left(\dfrac{1.962 \times 10^{-5}}{1.086}\right)^2 \times 322} \right] = 4.0568 \times 10^9$$

$Nu = 0.13 \times (Pr\,Gr)^{0.33} = 192.6$

$Nu = \dfrac{h.\ell}{k}$

$\therefore h_c = \dfrac{Nu.k}{\ell} = \dfrac{192.6 \times 2.816 \times 10^{-2}}{1} = 5.4$ W/m^{2}K

$Q_c = h_c A\ \Delta T = 5.8406 \times 1 \times (80-18) = 362$ W/m^2
$Q_R = \varepsilon\sigma A(T_s^4 - T_a^4) = 0.9 \times 5.67 \times 10^{-8} \times 1(353^4 - 291^4) = 426$ W/m^2

For ten radiators (two sides, $A = 20$ m^2) the total heat loss is:

$Q_t = 20(362 + 426) = 15.76$ kW

Worked example 3.7

A sports hall has the thermal specifications shown in Worked example 3.7, Table 1.

Worked example 3.7, Table 1

Item	U value (W/m²K)
Walls	0.5
Roof	0.45
Floor	0.45
Windows	3

The building is $20\,m \times 10\,m \times 5\,m$ high, and along its front wall it has windows with a total area of $12\,m^2$ and a door (U value, $2\,W/m^2K$, and area $6\,m^2$). Determine:

(a) The total fabric heat loss through this building when the inside temperature is kept at 20°C while the average outside temperature is 0°C.
(b) The energy cost due to heating for a typical season of 500 hours. Assume that the building is gas heated at the rate of 4 p/kWh.

Solution:

(a) It is very convenient to arrange the solution of such problems in a tabular form, as we have done in Worked example 3.7, Table 2.

Worked example 3.7, Table 2

Item	U value (W/m²K)	Area (m²)	Temperature difference	Heat loss (W)
Walls	0.5	282	20	2820
Roof	0.45	200	20	1800
Floor	0.45	200	20	1800
Windows	3	12	20	720
Door	2	6	20	240

Total fabric heat loss = 7.380 kW.

(b) Seasonal heating load = $7.38\,kW \times 500\,h = 3690\,kWh$
Cost of heating per season = $3690\,kWh \times 0.04\,£/kWh$
$$= £147.60$$

3.11 Tutorial problems

3.1 A composite wall is made up of external brickwork and a plaster board. The coefficients of thermal conductivity and thicknesses for these materials are as follows:

Brickwork $\quad\quad k = 0.60$ W/mK, $x = 110$ mm

Insulating board $\quad k = 0.06$ W/mK, $x = 25$ mm

The surface heat transfer coefficient on the inside wall is 2.0 W/m²K.

The surface heat transfer coefficient on the outside wall is 3.0 W/m²K.

Determine:

(a) The overall heat transfer coefficient for the wall.

(b) The heat loss through such a wall 6m high and 10m long.

Take the internal ambient temperature to be 20°C and the external ambient temperature to be 10°C.

Ans. (0.69 W/m²K, 418 W)

3.2 A wall is made up of 200mm brick (thermal conductivity 0.69 W/mK). If the inside and outside temperature and heat transfer coefficients are 20°C, 10°C, 20 W/m²K and 10 W/m²K respectively,

(a) Determine the steady state heat transfer per unit area through the wall.

(b) The wall in (a) was covered by a 10mm layer of plaster (thermal conductivity 0.05 W/mK) on both sides. Determine the saving in heating load by including the plaster.

Ans. (22.73 W/m², 47.6%)

3.3 A transformer is used for cooling oil at the rate of 50 kg/s. The oil enters at 60°C and leaves at 50°C. The oil is subsequently divided equally into ten tubes (internal tube diameter 10mm) in a heat exchanger. Calculate the internal heat convection coefficient using the empirical correlations. Use the following data: $\rho = 870$ kg/m³; $C_p = 2.0$ kJ/kgK; $\mu = 0.073$Pa s; Pr $= 1050$; $k = 140 \times 10^{-6}$ kW/mK.

Ans. (3686 W/m²K)

3.4 An electric heater consisting of a single horizontal bar with diameter 25 mm and length 0.5 m is used to maintain a room temperature of 17°C. If the heater's surface temperature is 537°C, calculate:

(a) The convective heat transfer.
(b) The radiative heat transfer if the heater surface has an emissivity of 0.3.

Use the following data:

Air properties:
$k = 2.816 \times 10^{-2}$ W/mK
$Pr = 0.701$
$\mu = 1.962 \times 10^{-5}$ kgm^{-1}s^{-1}
$\rho = 1.086$ kgm^{-3}

You may use the free convection from a vertical flat plate, for which the empirical relation is given by Nu = 0.13 × (Pr Gr)$^{0.33}$
Ans. (333 W, 282 W)

3.5 A central heating panel (with emissivity of 0.8) is placed by a brick wall (with emissivity of 0.9). The surface temperature of the heating panel is kept at 80°C while the brick surface is 20°C.

(a) Determine the radiation heat exchange per unit area between the two surfaces.
(b) In order to reduce the radiation between the two surfaces, a sheet of aluminium is placed in parallel between the two surfaces. If aluminium has an emissivity of 0.05, determine the heat exchange by radiation per unit area.

Ans. (340 W, 23 W)

3.6

(a) Calculate the heat loss through a wall (105 mm common brick, $k = 0.72$ W/mK; and 10mm plaster, $k = 0.6$ W/mK) with a total area of 100m² and containing a 10% area of single glazed windows (U value = 6 W/m²K) if the inside air and the outside design temperatures are 20°C and –1°C respectively. Assume $Ra_i = 0.123$ and $Ra_o = 0.055$ m²K/W.
(b) In order to reduce the heat loss of the building described above, the following two suggestions are to be considered:
    ● Double glazing (13 mm air cavity) the windows (U value = 3 W/m²K).
    ● Building an outer shell made from a 105mm common brick separated by a 50mm cavity filled with polystyrene ($k = 0.04$ W/mK).

Evaluate each scheme and comment on the results.
Ans. (6816 kW, the first suggestion gives a 9% improvement, while the second suggestion gives a 64% improvement and is therefore superior)

Chapter 4

Ventilation

Learning outcomes

- Describe the need for controlling air quality in occupied space. — Knowledge and understanding
- Describe the effect of occupancy on the composition of used air quality — Analysis
- Distinguish between natural and mechanical ventilation — Analysis
- Calculate the energy requirement for heating air — Analysis
- Demonstrate fan performance curves and fan selection procedure — Analysis
- Describe the process of CO_2 build-up in buildings — Knowledge and understanding
- Solve problems associated with ventilation — Problem solving
- Practise further tutorial problems — Problem solving

Energy Audits: A Workbook for Energy Management in Buildings, First Edition.
Tarik Al-Shemmeri.
© 2011 Blackwell Publishing Ltd. Published 2011 by Blackwell Publishing Ltd.

4.1 Aims of ventilation

The main purpose of ventilation is to maintain human comfort and health in buildings. To achieve this, a ventilation system should be able to:

- maintain adequate oxygen for life, or, in other words, control the CO_2 concentration inside an occupied zone;
- dilute odours produced continuously from human activity;
- control the concentration of airborne contamination such as fumes, dust, etc.;
- provide a sense of cooling by inducing a movement of the air.

The importance of ventilation becomes very significant when a problem arises. Health hazards attributed to improper space ventilation can be divided into four broad categories:

- *Physical*. Air pressure, heat, dampness, noise, radiant energy, electric shock.
- *Chemical*. Exposure to toxic materials such as dust, fumes and gases; dangerous levels of carbon monoxide and carbon dioxide (CO and CO_2); and VOCs (volatile organic compounds) released from carpets, furniture and building materials.
- *Biological*. Infections and viruses, e.g. tetanus, hepatitis, Legionnaire's disease and mould growth.
- *Ergonomic*. Work conditions and organisational stress, also related to sick building syndrome.

4.2 Air quality

Air consists of two main gases: nitrogen (78%) and oxygen (nearly 21% by volume), plus other gases present in much smaller volumes, as shown in Table 4.1.

In the ideal situation, pure air refers to a sample of air which is free from carbon dioxide and water vapour. These two gases are the products of human breathing and cooking associated with occupancy.

Clean air is essential for the wellbeing of humans; it has always been, and will continue to be, a requirement for healthy functioning of every person irrespective of their cultural, social or economic background. In 1952, pollution in London alone claimed 4000 lives over a short time period. This disaster was a lesson to the public and politicians. The Clean Air Act, passed by the United Kingdom Government in 1956, followed by a similar step by the US Government in 1963, brought air quality under the control of legislators in order to preserve health and safety of the public at large.

Enforcing the Clean Air Act resulted in a significant reduction in pollution-related fatalities. The World Health Organisation defines air pollution as:

Table 4.1 Composition of 'clean', dry atmospheric air.

Constituent	Molecular formula	Volume fraction
Nitrogen	N_2	78.084%
Oxygen	O_2	20.947%
Argon	Ar	0.934%
Carbon dioxide	CO_2	0.0314%
Neon	Ne	18.2 ppm
Helium	He	5.2 ppm
Methane	CH_4	1.5 ppm
Krypton	Kr	1.0 ppm
Hydrogen	H_2	0.5 ppm
Nitrous oxide	N_2O	0.25 ppm
Carbon monoxide	CO	0.1 ppm
Ozone	O_3	0.02 ppm
Sulphur dioxide	SO_2	0.001 ppm
Nitrogen dioxide	NO_2	0.1 ppm

ppm = parts per million.

'The presence in the air of one or more contaminants, such as dust, fumes, gas, mist, odour, smoke or vapour, in quantities or characteristics, and of duration such as to be injurious to human, plant or animal life or to property, or which unreasonably interferes with the comfortable enjoyment of life and property.'

Nowadays, increasing recognition is given to the great proportion of time, perhaps as much as 70-80%, spent indoors. Therefore, it is imperative that the air quality is maintained to the best possible standards.

4.2.1 Minimum fresh air requirements

The fresh air quantity required merely to provide sufficient oxygen is small, and so fresh air requirements (usually referred to as *ventilation rates*) are based on the need to dilute the carbon dioxide and odours emanating from the occupants.

The European CEN Standards specify ventilation rates in the range of 15-25 L/s for each occupant present in a space. The UK Building Regulations (Part F 2006) specify a minimum ventilation rate for office premises of 10 L/s per person.

4.2.2 Composition of respired air

The relative proportion of water vapour existing in the atmosphere is increased by contact of the air with water and by the perspiration and breathing of occupants in confined spaces. At the same time, the presence of carbon dioxide existing in the atmosphere is increased whenever oxidising processes are encountered, such as in fuel burning and breathing in humans. The exact composition of air varies with type and length of occupancy, and typical measurements in a confined exercise room are shown in Table 4.2.

A fall in oxygen level from a normal value of 21% (by volume) to as little as 13% escapes notice. An increase in CO_2 level from a normal value of 0.03% to 2% increases the depth of respiration by 30%. The *threshold limit value (TLV)* may be defined as the airborne concentration of a substance that represents conditions to which it is believed most working people may be exposed without

Table 4.2 Composition of typical respired air.

Constituent	Atmospheric air percentage by volume	Respired air percentage by volume
Nitrogen	78.00	75.00
Oxygen	20.26	16.20
Water vapour	1.70	4.80
Carbon dioxide	0.04	4.00

Table 4.3 Recommended winter indoor temperatures for common buildings.

Type of room	Winter indoor temperature °C
Bar	18
Bathroom	22
Bedroom	18
Classroom	20
Corridor	16
Dining room	20
Factory, sedentary work	18
Factory, light work	16
Factory, heavy work	13
Gym	15
Hotel room	21
Laboratory	20
Lecture room	20
Library	20
Living room	21
Museum	20
Office	20
Recreation room	18
Restaurant	18
Shop	18
Store	15
Swimming baths	27

Adapted from http://www.engineeringtoolbox.com/

Table 4.4 Recommended air changes for common buildings.

Building/room	Air change rates per hour - N -
Auditorium	8–15
Bank	4–10
Bar	20–30
Bowling alley	10–15
Cafeteria	12–15
Church	8–15
Department store	6–10
Dining hall	12–15
Dining room, hotel	5
Factory building, fumes etc.	10–15
Kitchen	15–60
Library, public	4
Nightclub	20–30
Office, public	3
Office, private	4
Police station	4–10
Post office	4–10
Precision manufacturing	10–50
Pump room	5
Restaurant	8–12
Retail	6–10
School classroom	4–12
Supermarket	4–10
Warehouse	2
Waiting room, public	4

Adapted from http://www.engineeringtoolbox.com/

adverse effect. The TLV for an eight-hour exposure to CO_2 is 5% and it is generally believed that the upper acceptable level is 0.1%.

The most dangerous pollutant from smoking is carbon monoxide (CO), which has a TLV for an eight-hour exposure of 66 ppm. However, a minority of people would be affected by values lower than this and the American recommendations are for a figure of 9 ppm (requiring a dilution rate of 9 m^3 per cigarette!).

Tables 4.3 and 4.4 indicate the standard recommended indoor temperatures and ventilation requirements for typical buildings.

4.3 Ventilation methods

For air to move into and out of a building, a pressure difference between the inside and outside of the building is required. The resistance to flow of air through the building will affect the actual air flow rate. This movement of air through the building can be created either by relying on natural means to provide the outside air or by forcing it mechanically inside the building. It is obvious that, whenever possible, if occupancy is not so dense, the natural option should be chosen as this requires no energy input.

4.3.1 Natural ventilation

Natural ventilation has historically been the means of allowing fresh air from outside to enter an occupied space. This can be done in one of two ways, or by a combination of the two:

- *Controlled natural ventilation*. This is intentional displacement of air through specified openings such as windows and doors, and ventilation by using natural forces (usually pressure from wind and/or indoor-outdoor temperature differences).
- *Infiltration*. This is the uncontrolled, random flow of air through unintentional openings driven by wind, temperature-difference pressures and/or appliance-induced pressures across the building envelope.

In general, controlled natural ventilation and infiltration are driven by pressure differences across the building envelope. These pressure differences are caused by:

- wind (wind effect);
- a difference in air density due to a temperature difference between the indoor and outdoor air (stack or chimney effect);
- a combination of both wind and stack effects.

Wind effect-driven natural ventilation depends on three parameters:

- wind speed and direction;
- location of the building and surrounding topology;
- shape, size and orientation of the building.

The relationship between the pressure difference and air motion (V_w) may be expressed as a modified Bernoulli theorem:

$$P_w - P_o = C_p \cdot \frac{1}{2} \cdot \rho \cdot v_w^2 \qquad\qquad [4.1]$$

The stack effect relies on the fact that the pressure difference between the interior and exterior is related to the height (h) from the neutral pressure level and the difference between the densities of inside and outside air. The stack effect is governed by the hydrostatic equation: $\Delta P = \rho g \Delta h$.

Since both the density and the change in elevation vary in different parts of the building, the stack effect can be written in a more general form:

$$\Delta P_s = (\rho_o - \rho_i) \cdot g \cdot (h - h_{neutral}) = \rho_i \cdot g \cdot (h - h_{neutral}) \cdot \frac{T_i - T_o}{T_o} \qquad [4.2]$$

where $h_{neutral}$ is the height of the neutral pressure level (m) near the centre point, and T = absolute temperature (K) (subscripts: i = inside; o = outside).

4.3.2 Mechanical or forced ventilation

Adequate ventilation is never guaranteed with naturally induced ventilation, as it depends on the forces of nature to provide it. For limited or low occupancy it is possible to get away with it, but in offices, supermarkets, theatres or cinemas, where occupancy is heavy, and where comfort is paramount, forced ventilation is used to ensure that adequate good air quality is maintained.

There are three methods or systems to provide the required ventilation (Figure 4.1):

(1) *Extract systems*. Where the air is vented by fans and replacement air enters naturally at ground level, either through the building structure or correctly positioned air inlet louvres. This method is particularly good for localised control, such as kitchens, toilets, and so on. It relies on the natural induction of fresh air, but the fan will take away polluted air from points of high concentration.

(2) *Supply systems*. Where the air is supplied by a fan-powered system. In this case, the system can be combined with a heating system. This system has the advantage of introducing fresh or treated air to points of particular interest, such as living rooms, etc. It relies on the fact that an open window at a higher level will allow the warmer air to leave.

(3) *Supply and extract systems (also known as mechanically balanced systems)*. A combination of the above, where air is supplied and extracted by fan-powered systems to give a better distribution. This system is very efficient in making sure that adequate ventilation is guaranteed, however this comes at a cost - two fans consume double the amount of energy used in the previous cases.

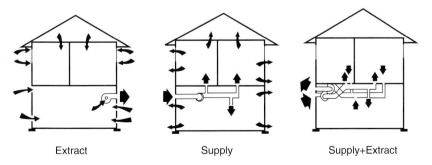

| Extract | Supply | Supply+Extract |

Figure 4.1 Forced ventilation systems.

The decision to choose any of the above methods requires careful examination of the environmental and economic aspects.

4.4 Ventilation flow calculations

Generally, figures for volume flow rate, pressure drop and energy losses are needed in ventilation calculations.

4.4.1 Volume flow calculations

The volume flow rate, V_f (m³/s) of ventilation can be calculated using the volume of the building, Vol, and the number of air changes per hour, n, as:

$$V_f = Vol \times \frac{n}{3600} \qquad [4.3]$$

If the velocity of flow (v) in a conduit of cross-sectional area A is known, then the volumetric flow rate of supply air (V_f) and the mass flow rate (m_f) can be calculated as:

$$V_f = v \times A; \quad m_a = V_f \times \rho \qquad [4.4]$$

4.4.2 Ventilation heat load calculations

The ventilation rate required to remove heat (Q_v) from an occupied space is given by the product of the mass of supply air (m_a), the specific heat capacity of air (Cp) and the temperature rise (ΔT):

$$Q_v = m_a \, Cp \, \Delta T \qquad [4.5]$$

In addition, the term 'air changes per hour' (n) is also commonly used, so the ventilation heating load is written as:

$$Q_v = 0.335 \, n \,.Vol. \, \Delta T \qquad [4.6]$$

The constant (0.335) is derived from the specific heat capacity of air, on average equal to 1005 J/kgK, and the average air density of 1.2 kg/m³.

4.4.3 Ventilation calculations based on CO_2 build-up

Indoor air quality is affected by the type of occupancy and the quality and quantity of air supplied to the space. The amount of CO_2 inside a building increases with time, and the final CO_2 concentration (C_t) any time, t, can be predicted in terms of the initial concentration at time zero, and the type of occupancy (production rate, P, and volume flow rate of supply air V_f and the number of air changes per hour, n):

$$C_t = ((C_o + 10^6 \times P/V_f) \times (1 - e^{-nt})) + C_i \times e^{-nt} \qquad [4.7]$$

where C is the concentration of CO_2 (ppm); with suffixes t = in hours after time t_i = initial (at time = 0), o = outdoor air. P is the rate of release of CO_2 (for example in m^3/s), V_f is the outdoor air supply rate (in the same units as P, e.g. m^3/s).

In order to estimate the quantity of air (V_f) that needs to be changed on a regular basis for the comfort of people inside a building, and to release the build-up of CO_2 indoors, knowing the concentration of CO_2 for the supplied air, C_s, and the threshold concentration of CO_2 recommended for a given occupancy, C_r and using the mass flow balance across the room implies that:

$$P = V_f \times C_r - V_f \times C_s$$

Hence

$$V_f = \frac{P}{C_r - C_s} \ (m^3/s) \qquad [4.8]$$

For example, for a sedentary adult, the following figures are recommended:

$$P = 4.70 \times 10^{-6} \ (m^3/s); \quad C_s = 0.03\%; \quad C_r = 0.50\%$$

Hence

$$V_f = \frac{P}{C_r - C_s} = \frac{4.70 \times 10^{-6}}{(0.50 - 0.03) \times 10^{-2}} = 1.0 L/s$$

The 1.0 L/s ventilation rate derived above is based on the carbon dioxide exhaled by one person engaged in a sedentary activity. Other important factors to be considered include: body latent heat, body odour and fumes (from smoking) if applicable.

4.5   Fans

In order to control the rate of ventilation, fans are used. Three main types of fan are used in air-conditioning systems:

- *Propeller fans*. This type is usually used at free openings in walls or windows and for other types of low pressure applications. It is not usually employed for ducted ventilation systems.
- *Axial-flow fans*. This type of fan is designed for mounting inside a duct system and is suitable for moving air in complete systems of ductwork.
- *Centrifugal fans*. The inlet of the fan is at 90° to the outlet and, like the axial-flow fan, it is suitable for moving air in complete systems of ductwork.

4.5.1 Fan laws

The performance of a fan incorporated in a system of ventilation is governed by the following laws, assuming the air density remains constant:

Flow rate law $\quad \dfrac{V_{f1}}{V_{f2}} = \dfrac{N_2}{N_1}$ [4.9]

where V_f is the volumetric flow rate and N is the rotational speed of the fan.

Pressure rise law $\quad \dfrac{\Delta P_2}{\Delta P_1} = \dfrac{(N_2)^2}{(N_1)^2}$ [4.10]

where ΔP is the pressure across the fan.

Fan power law $\quad \dfrac{E_2}{E_1} = \dfrac{(N_2)^2}{(N_1)^2}$ [4.11]

where E is the energy absorbed by the fan.

The efficiency of a fan can be found by using the following expression:

Efficiency, $\eta = \dfrac{\Delta P \times V_f}{E} \times 100\%$ [4.12]

4.5.2 Selection of fans

In order to select a fan (Figure 4.2) for a given purpose, reference should be made to fan performance graphs supplied by the manufacturer. The fan should

Figure 4.2   Selection of fans. Courtesy of Soler & Palau Ltd.

be chosen to match the application, and this is usually done by calculating the system duty, then choosing a fan with a rated power higher than the calculated one in order to allow for a drop in efficiency afterwards.

4.5.3 Calculation of ventilation fan duty

Many fan systems are not run at their maximum efficiency. Dampers, throttling valves, by-pass systems and pressure relief valves are often used to reduce output to a level that matches demand. The motor for the ventilation system has to cater for the load of moving the air from outside to the place of application; in addition, it has to cope with the frictional resistance due the ducting surfaces, controls and fittings. All of these are grouped into one term – PRV, the pressure resistance.

Fans' and blowers' energy input is proportional to three variables:

V_f = Volume flow rate
ΔP = Pressure resistance of the ventilation system
Eff = Efficiency of the driving motor

The general equation for calculating the power required is given by:

$$\text{Power} = \frac{V_f \times \Delta P}{Eff} \qquad [4.13]$$

where

$$PRV = \Delta P_{friction} + \Delta P_{obstructions} + \Delta P_{static} \qquad [4.14]$$

$\Delta P_{friction}$ = the pressure drop due to friction resistance to the flow, by Darcy equation, see the next section for calculations.

$\Delta P_{obstructions}$ = the pressure drop due to a number of obstructions such as bends, filters, valves, etc., usually quoted as $\Sigma k \left(\rho \dfrac{v^2}{2} \right)$. Σk represents the sum of all coefficients of fittings in the duct. k values can be found from standard fluid books or from manufacturers of ducts.

ΔP_{static} = due to the difference in elevation between suction and delivery point, quoted as $\Delta P_{static} = \rho g h_s$ where h_s is the difference in height.

4.5.4 Pressure drop calculation

The pressure drop due to friction resistance to the flow in a duct is given by the Darcy equation:

$$\Delta P = \frac{4 f L}{De} \rho \frac{v^2}{2} \qquad [4.15]$$

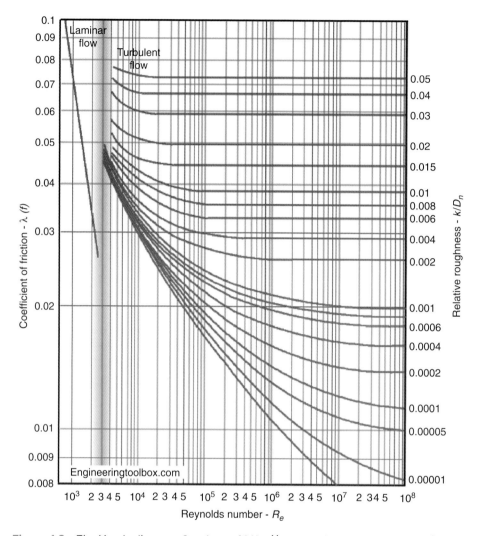

Figure 4.3 The Moody diagram. Courtesy of http://www.engineeringtoolbox.com/

where
L is the length of the duct (m)
De is the diameter (if circular) or equivalent diameter for noncircular cross-sections calculated by the ratio: 4 × cross-sectional area divided by the perimeter.
ρ is the density of air (1.2 kg/m³ for ambient air can be assumed)
v is the velocity of flow (m/s)
f is the friction factor, which can be either taken from the Moody diagram (Figure 4.3) or calculated for smooth ducts by:

$$f = \frac{16}{Re} \text{ if } Re < 2000 \quad \text{and} \quad f = \frac{0.079}{Re^{.25}} \text{ if } Re \geq 2000 \qquad [4.16]$$

Re is a dimensionless quantity known as the Reynolds number, and is defined as:

$$Re = \frac{\rho \, v \, De}{\mu} \qquad\qquad [4.17]$$

where μ is the dynamic viscosity of air.

4.5.5 Energy efficiency in ventilation systems

Increasing efficiency may involve a combination of actions: trimming impellers, installing variable speed drives, improving controls or improving the efficiency of motors. Other benefits include the extension of equipment life, a reduction in safety hazards and a reduction in noise and vibration. A unit of energy saved at the fan saves three units at the motor, so it makes sense to start saving energy at the end of the system and work backwards. The following procedure can actually help in optimising energy consumption.

(1) Choose the right sized ventilation system.

- *Fan size.* It is common for most building fan systems to be oversized by an average of 60%. This represents an enormous amount of wasted energy. By appropriately resizing the fan system, energy costs can be saved.
- *Motor size.* Replacing old motors with appropriately sized, highly efficient ones will save large amounts of energy.
- *Pipe and duct size.* Capital saved by installing undersized piping or ductwork can be a very costly false saving, as it puts additional loads on the fans and motors. A 15% increase in pipe diameter can cut pressure drop in half. This allows the fan, motor and consequently energy consumption to be much smaller. Minimising the number of bends and valves in pipework will reduce energy lost to friction.

(2) Check the fan system inefficiency.

- Check the system to find out if any throttling valves or dampers are constantly restricting the rate of flow to 10% less than its design flow rate. If so, consider impeller trimming, as this reduces energy demand.
- Noise and vibration may indicate efficiency problems.

(3) Control the controls.

- *Turn it off.* Turn off the fan or pump when it is not needed. This can be done manually or automatically with the installation of controls.
- *Slow it down.* The most effective way to match output to demand and to save energy is to reduce the speed of load (i.e. the pump or fan). Reducing a fan's speed by 20% can reduce its energy requirements by nearly 50%.

(4) Establish a regular maintenance programme.

- Clean pumps and fans to ensure maximum efficiency. Accumulation of dirt or dust can decrease the efficiency of a fan by adding weight to it and increasing pressure loss in the system. Important components to clean are the filters, heating coils, silencers and fan blades.
- Replace worn seals and fan blades.
- Use low friction coatings on the internal surfaces of pumps to improve pump efficiency.
- Ensure that drive belts are in good condition, evenly matched and correctly sized.

4.6 Worked examples

Worked example 4.1

(a) A ventilation duct with a 1m diameter circular cross-section carries the fumes from a large office. If a rate of two air changes per hour is necessary, determine the capacity (m³/minute) of the fan required for this duty. The room size is 10 m × 6 m × 3 m.

(b) It has been decided to replace the above duct, and the only replacement available is a square duct with a side measuring 0.25 m. Determine the velocity for this duct if the above ventilation capacity is to be maintained. Comment on the new result.

Solution:

(a) Volume flow rate (m³/min) = number of air changes per hour × volume of space/60

$$V_f = 2 \times (10 \times 6 \times 3)/60 = 6\,m^3/min$$

$$\text{Velocity in the duct} = \frac{\text{Volume flow}}{\text{Cross-sectional area}} = \frac{6/60}{\frac{\pi}{4} \times 1^2} = 0.127\,m/s$$

(b) For the square duct:

$$\text{Velocity in the duct} = \frac{\text{Volume flow}}{\text{Cross-sectional area}} = \frac{6/60}{0.25 \times 0.25} = 1.6\,m/s$$

Luckily this velocity is lower than the maximum velocity recommended to limit noise, $V_{max} = 5\,m/s$.

....

Worked example 4.2

An office measuring $10 \times 10 \times 3\,m$ high is to be ventilated at the rate of five air changes per hour and with an air flow rate of $4\,m/s$ in the supply duct. Determine:

(a) The volume flow rate.
(b) The dimensions of a square or circular duct for the supply air line.

Solution:

(a) The volumetric flow rate is $V_f =$ volume of building $\times$ air changes per second

$$= (10 \times 10 \times 3) \times \frac{5}{3600} = 0.417\,m^3/s$$

(b) In the second definition, volumetric flow is given by: $V_f =$ area of duct $\times$ velocity of flow

Hence, the area of cross-section is calculated as follows:

$$A = \frac{V_f}{V} = \frac{0.417}{4} = 0.104\,m^2$$

The area of a square duct is:

Area $= X^2$

Hence, $X = 0.322\,m$
For a circular duct, the diameter is found as:

$$A = \frac{\pi}{4}D^2$$

$$\text{Hence } D = \sqrt{\frac{A}{\pi/4}} = \sqrt{\frac{0.104}{\pi/4}} = 0.363\,m$$

which is larger than a square duct.

....

Worked example 4.3

A gym has dimensions of $30 \times 20 \times 4\,m$ high, and is occupied by a maximum of 30 people. The ventilation system operates on supply rates of $12\,L/s$ fresh air and $48\,L/s$ of recirculated air per person. Calculate the building air changes per hour.

Solution:

The total volume of air supplied is the sum of recirculated and fresh air:

$$V_t = (12 + 48) \times 30 = 1800 \text{L/s} = 1.8 \text{m}^3/\text{s}$$

By definition, the volume flow is equal to the product of building volume and the rate of air changes:

$$V_t = \text{Volume} \times \frac{n}{3600}$$

Rewriting the above equation, the number of air changes required can be calculated as follows:

$$n = \frac{3600 \times V_t}{\text{volume}}$$
$$= \frac{3600 \times 1.8}{30 \times 20 \times 4}$$
$$= 2.7$$

Worked example 4.4

Ten people exercise simultaneously in a gymnasium. If the outdoor air quality has a CO_2 concentration of 0.05%, determine the rate of air supply necessary to avoid a build-up of CO_2 such that the concentration rises above the maximum limit of 0.25%. Assume the production rate for each person is $10 \times 10^{-6} \text{m}^3/\text{s}$.

The following data are known:

Production rate of CO_2, $P = 10 \times 10^{-6}$ (m³/s)
Concentration upon leaving, $C_r = 0.25\%$
Concentration when entering, $C_s = 0.05\%$

Solution:

The ventilation rate is determined by using the following equation:

$$V_f = (P)/(C_r - C_s)$$

Hence

$$V_f = \frac{10 \times 10^{-6}}{(0.25 - 0.05) \times 10^{-2}}$$
$$= 0.005 \text{m}^3/\text{s per person}$$

For ten people, the value will be 0.05 m³/s.

Worked example 4.5

The supply duct from an air-conditioning system has a flow rate of $2\,m^3$/min, which in turn splits (not equally) into two ducts of 10 cm × 10cm in Room 1 and 15 cm × 15 cm in Room 2 (see Worked example 4.5, Figure 1). If the air velocity in the main duct is 0.5 m/s, and that in the duct leading to Room 1 is 2 m/s, determine, for an incompressible flow:

(a) The main duct size.
(b) The air velocity in the duct leading to Room 2.
(c) The required size of duct necessary to halve the velocity in Room 1.

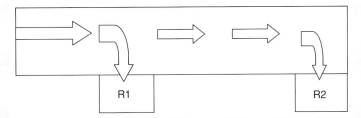

R1   R2

Worked example 4.5, Figure 1

Solution:

(a) Volume flow rate = area × velocity in the main duct

$$\frac{2}{60} = \frac{D}{100} \times \frac{D}{100} \times 0.5 \qquad \therefore D = 25.82\,\text{cm}$$

(b) Volume flow from the main duct divides into two branches, hence:

$$\frac{2}{60} = A_1 V_1 + A_2 V_2$$

$$\frac{2}{60} = \frac{10}{100} \times \frac{10}{100} \times 2 + \frac{15}{100} \times \frac{15}{100} \times V_2$$

$$0.0333 = 0.02 + 0.0225\, V_2 \qquad \therefore V_2 = 0.59\,\text{m/s}$$

(c) Volume flow rate = area × velocity in duct 1.

$$0.02 = D_1^2 \times 1$$
$$\therefore D_1 = 14\,\text{cm}$$

Worked example 4.6

Consider an office with dimensions 10 × 6 × 3 m high which lies unused all weekend. At 9am on Monday morning 30 adults, each exhaling 0.005L/s of CO_2 enter the office where the fresh air (at 400 ppm) ventilation rate is two air changes per hour. Calculate the predicted CO_2 level at midday.

Solution:

$t = 3$ hours (midday 12pm, starting time 9am)
$n = 2$ air changes per hour
$C_o = 400$
$P = 30 \times 0.005 = 0.15\,L/s$
$V_f = \text{vol} \times n / 3600$
$\quad = (10 \times 6 \times 3) \times 2/3600$
$\quad = 0.1\,m^3/s$

Use the following expression:

$C_t = ((C_o + 10^6 \times P/V_f) \times (1 - e^{-nt})) + C_i \times e^{-nt}$
$CO_2 \text{ at midday} = [(400 + (10^6 \times 0.15/100)) \times (1 - e^{-(2\times3)})]$
$\quad\quad\quad + (400 \times e^{-(2\times3)})$
$\quad\quad\quad = [(400 + (10^6 \times 0.0015)) \times (1 - 0.0025)]$
$\quad\quad\quad + (400 \times 0.0025)$
$\quad\quad\quad = 1900\,ppm$

Worked example 4.7

A fume extraction system has two extraction hoods, each connected to the main duct by rectangular ducts measuring 300 mm by 450 mm and with lengths 30 m and 40 m respectively. The main duct is also rectangular and measures 400 mm by 600 mm, with a length of 25 m. It is proposed that a fan with the characteristics outlined in Worked example 4.7, Table 1 be used to operate the system.

Worked example 4.7, Table 1

Flow rate (m³/s)	0	1.0	1.5	2.0	2.5	3.0	3.5	4.0
Pressure rise (KN/m²)	0.15	0.50	0.62	0.72	0.75	0.68	0.60	0.48
Overall efficiency (%)		50	85	90	86	82	75	64	55
System pressure drop (Pa)								

Assume that the fluid friction factor f for the ducts remains constant at 0.012 and the density of air circulated is 1.2 kg/m³. Determine:

(a) The total quantity of air extracted by the fan.
(b) The fan shaft power.
(c) The daily operating cost over a 10-hour shift if electricity costs 7p per kWh.

Solution:

The pressure drop due to friction is:

$$\Delta P = \frac{4fL}{De} \times \frac{\rho v^2}{2} = Rv^2$$

where

$$R = \frac{4fL}{De} \times \frac{\rho}{2}$$

The equivalent diameter

$$De = \frac{4 \times Area}{Perimeter}$$

$$De_{main} = \frac{4(0.400 \times 0.600)}{2(0.400 + 0.600)} = 0.48 \text{ m}; \quad R_{main} = \frac{4 \times 0.012 \times 25 \times 1.2}{2 \times 0.48} = 1.5$$

$$De_{30} = \frac{4(0.300 \times 0.450)}{2(0.300 + 0.450)} = 0.36 \text{ m}; \quad R_{30} = \frac{4 \times 0.012 \times 30 \times 1.2}{2 \times 0.36} = 2.4$$

$$De_{40} = \frac{4(0.300 \times 0.450)}{2(0.300 + 0.450)} = 0.36 \text{ m}; \quad R_{40} = \frac{4 \times 0.012 \times 40 \times 1.2}{2 \times 0.36} = 3.2$$

Take the largest resistance route, i.e. 1.5+3.2=4.7 units of resistance.

$$\Delta P = R \, v^2 = 4.7 \, v^2$$

This equation is used to calculate the system pressure drop at various flow rates (which are converted into velocity by dividing by the main duct area).

Thus, the table is completed.

V_f (m³/s)	1.0	1.5	2.0	2.5	3.0	3.5
ΔP (kN/m²)	0.06	0.14	0.26	0.40	0.58	0.78

Draw the performance plot (Worked example 4.7, Figure 1), to determine the operating point, at 3.2 m³/s, ΔP = 0.65 kN/m² and efficiency = 73%.

The power consumption is:

$$E = \frac{V_f \times \Delta P}{\eta} = \frac{0.65 \times 3.2}{0.73} = 2.85 \, \text{kW}$$

Daily energy cost = consumption × duration × cost of unit consumed:
$$= 2.85 \times 10 \times 7 = £1.99$$

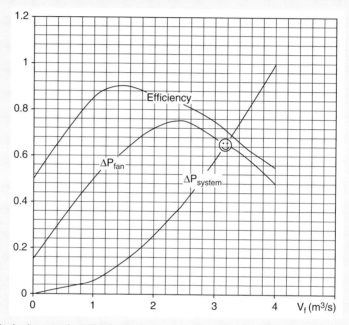

Worked example 4.7, Figure 1

Worked example 4.8

A ventilation system for a swimming pool consists of an extraction hood connected to the main duct by a rectangular duct measuring 300 mm by 300 mm and with length 20 m. The duct surface has a friction factor of 0.005 and a total obstruction losses factor of 5; the velocity of air flow in the duct is limited to 8 m/s. There are two fans to choose from (Fan A and Fan B, shown graphically in Worked example 4.8, Figure 1) and both fans cost the same. Select the most suitable fan, justifying your choice.

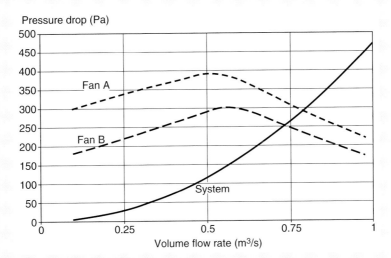

Worked example 4.8, Figure 1

Solution:

The system total load (pressure difference due to components' frictional losses) is calculated as the sum of the friction and total obstruction fittings:

$$\Delta P_{system} = \left(5 + \frac{4fL}{De}\right) \times \frac{1}{2} \times \rho V^2 = \left(5 + \frac{4 \times 0.005 \times 20}{0.3}\right) \times \frac{1}{2} \times 1.2 \times \left[\frac{V_{fsys}}{0.3 \times 0.3}\right]^2$$

$$= 469 V_{fsys}^2$$

ΔP_{system} (Pa)	4.7	18.7	42.2	75	117	169	230	300	380	469
Flow rate (m³/s)	0.1	0.2	0.3	0.4	0.5	0.6	0.7	0.8	0.9	1.0

At the intersections (the operating points), Fan A will provide a volume flow rate of 0.8 m³/s with a pressure drop of 290 Pa, while Fan B will provide 0.70 m³/s at 250 Pa. Thus, Fan A will be the better of the two, since it will ensure the required volume flow rate of 0.72 m³/s. Fan B would deliver a flow rate slightly lower than the designed rate.

Worked example 4.9

A fan absorbs 2.3 kW of power and discharges 2.5 m³/s when the impeller angular velocity is 1000 revolutions per minute. If the impeller angular velocity is increased to 1200 revolutions per minute, calculate the discharge in m³/s and the power absorbed for this new condition.

Solution:

Using the Flow rate fan law:

$$\frac{V_{f2}}{V_{f1}} = \frac{N_2}{N_1}$$

$$\therefore V_{f2} = V_{f1} \times \frac{N_2}{N_1}$$

$$= \frac{2.5 \times 1200}{1000} = 3 \, m^3/s$$

Using the Fan power law:

$$E_2 = \frac{E_1 \times (N_2)^3}{(N_1)^3}$$

$$= \frac{2.3 \times 1200^3}{1000^3} = 3.974 \, kW$$

Worked example 4.10

(a) A factory building measures 40 m × 20 m × 7 m. If the outside air supply temperature is 10°C, determine the ventilation heat load for this building. Assume that a comfortable temperature for the occupants is 17°C and that the recommended ACH is 3.

(b) The factory moved out and the building is now used as a warehouse. Determine the energy saving possible if the heating system is adjusted to the new occupancy (T = 18 and ACH = 2) while the supply air temperature is assumed the same as in (a).

Solution:

$$V = 40 \times 20 \times 7 = 5600 \, m^3$$

The air change load is calculated by the following equation:

$$Q_v = 0.335 \ N \ V \ \Delta T$$

(a) For the factory, $T = 17$ and ACH = 3. Hence, the ventilation heat loss is calculated by:

$$Q_v = 0.335 \times 3 \times 5600 \times (17 - 10)$$
$$= 39.396 \ kW$$

(b) For a warehouse, $T = 18$ and ACH = 2. Hence, the ventilation heat loss is calculated by:

$$Q_v = 0.335 \times 2 \times 5600 \times (18 - 10)$$
$$= 30.016 \ kW$$

The energy saving is 24%.

4.7 Tutorial problems

4.1   A ventilation duct with a 1.2 m diameter circular cross-section carries the fumes from a large hall. If a rate of 1.5 air changes per hour is necessary, determine the capacity (m³/min) of the fan required for this duty. The room size is 12 m × 10 m × 5 m.
Ans. (15 m³/min, 0.22 m/s)

4.2   An office measuring 10 × 10 × 4 m high is to be ventilated at a rate of three air changes per hour. If the air flow in the supply duct is limited to a rate of 4 m/s, determine:

(a) The volume flow rate.
(b) The dimensions of a square duct for air supply.

Ans. (0.333 m³/s, 0.288 m)

4.3   A gym has dimensions 30 × 20 × 4 m high and is occupied by a maximum of ten people. The ventilation system operates on supply rates of 12 L/s fresh air and 48 L/s of recirculated air per person. Calculate the number of air changes per hour.
Ans. (0.6 m³/s, 0.9)

4.4 Twenty people exercise simultaneously in a gymnasium. If the outdoor air has a CO_2 concentration of 0.07%, determine the rate of air supply necessary to avoid the build-up of CO_2 such that the concentration does not exceed the maximum limit of 0.25%. Assume the production rate for each person is $10.4 \times 10^{-6}\,m^3/s$.
Ans. ($0.1115\,m^3/s$)

4.5 A fan absorbs 3 kW of power and discharges $2.9\,m^3/s$ when the impeller angular velocity is 1000 revolutions per minute. If the impeller angular velocity is increased to 1200 revolutions per minute, calculate the discharge in m^3/s and the power absorbed for this new condition.
Ans. ($3.48\,m^3/s$, 5.184 kW)

4.6 A fan develops a static pressure of 300 Pa when the angular velocity of the impeller is 800 revolutions per minute. If the angular velocity of the impeller is increased to 1000 revolutions per minute, calculate the static pressure developed by the fan for the new condition. Determine the efficiency of the original system if its volume flow rate is $5\,m^3/s$ and the power absorbed is 2.0 kW.
Ans. (468.75 Pa, 75%)

4.8 Case Study: The National Trust's ventilation system

In this section a natural ventilation system is selected to demonstrate the huge saving that can be made on energy consumption costs as well as the associated reduction in CO_2 whilst maintaining environmental comfort within the building.

secontrols.com

Case Study
The National Trust Central Office

Name:
Heelis, The National Trust Central Office

Location:
Swindon

Title:
Providing a highly sustainable ventilation strategy, one that maintains a comfortable internal temperature for the National Trust's 430 staff, cuts CO_2 emissions and keeps running costs to a minimum.

Challenge:
To design, manufacture and commission a natural ventilation system for what is probably the most sustainable, low energy office development in the UK.

Products:
- OSO control system
- 307 actuators installed on 213 windows
- 156 actuators installed on 78 roof vents

Benefits:
Running costs to be reduced by £550,000 a year, with further savings of over £650,000 expected thanks to improved internal working practices and London weighting payments.

The building itself is expected to generate only 21kg of CO_2 per square metre annually, compared to 169kg for a typical air conditioned office. It has also won a number of industry accolades as a result, including a prestigious RIBA Sustainability Award.

"SE Controls was chosen because of the transparency of their design and the system's ability to work and communicate easily with the BMS."
Guy Nevill
Max Fordham LLP

The National Trust, as a charity, works to preserve and protect the coastline, countryside and buildings of England, Wales and Northern Ireland.

It currently opens to the public over 300 historic houses and gardens, and 49 industrial monuments and mills – as well as preserving a vast number of forests, woods, fens, beaches, farmland, downs, moorland, islands,

archaeological remains, castles, nature reserves and villages.

The Trust invests over £160 million a year into the nation's environmental infrastructure and works with over 40,000 companies, including 2,000 specialist conservation businesses.

For an organisation so dedicated to the ecological challenges of our time, it was imperative that The Trust's new

conducive to interactive team working between departments.

The benchmark of green building

The resulting building, headquarters named Heelis, was created by developers Kier Ventures and architects Feilden Clegg Bradley in collaboration with The National Trust, natural ventilation specialists SE Controls / NVS and Max Fordham LLP .

Cutting edge in its design, Heelis beautifully complements the surviving buildings of the Swindon Railway Works with their distinctive facades by Brunel and Gooch. Internally, the striking structure has a generous front-of-house atrium with a shop and café that is open to the public. It also comprises a two-storey deep open plan office.

Two courtyards were included in the scheme, introduced by Max Fordham LLP and the architect, to enable cross ventilation. In such a deep space, the courtyards both required vents leading into them – without which the cross ventilation strategy would not be able to function effectively.

A number of 'green' products and systems were also specified for the project, including a bank of 1,554 photovoltaic solar panels on the roof. Electricity generated from these panels is used in the building. A lighting control system adjusts the level of artificial light in response to external conditions and movement sensors ensure lights are turned off in unoccupied areas.

All of the timber used in construction has been harvested from sustainable woodland, much of it originating from National Trust properties. Even the carpet was specially developed using wool from Herdwick sheep grazed on National Trust farmland.

Blue Staffordshire engineering bricks were used as the principal external material, blending in as a contemporary interpretation of the surrounding structures. They are laid in lime mortar to reduce cement use and facilitate recycling. Cast aluminium cladding (visible on parts of the elevation) also provides a subtle tribute to the railway works that once occupied the site.

headquarters building would possess all the same sustainable and responsible credentials as its work ethos.

Inherent sustainability

With staff spread across four different sites, some teams having three or four locations, The National Trust needed a new central office – one that could house everyone under one roof and provide its members with a better, more integrated service. As well as streamlining working practices and cross-team communication, The Trust also wanted to reduce costs and pollution with a smaller carbon footprint.

Rather than choosing a typical country house, The National Trust sought better value for money in the former railway town of Swindon. It chose to occupy a large brownfield site in the Great

'The success of our sustainable initiatives is down to managing expectations. Every member of staff has had training on the natural ventilation system.'

Liz Adams
National Trust Property Manager

Western Railway Works, an area of significant architectural heritage undergoing major regeneration.

Sustainability was a key issue for The Trust and its move to Swindon. As such, the new build had to meet stringent and high quality benchmarks for sustainable design – as well as providing an open plan space

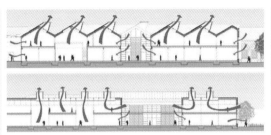

Image courtesy of Max Fordham LLP

Ensuring optimum internal temperatures

Internally, the building required a ventilation strategy that would provide fresh air and generate a healthy environment for its users. It also needed an alternative method of cooling other than air conditioning – one of the largest producers of co_2 emissions and incredibly expensive to run. In line with the building's myriad sustainability credentials, air conditioning and mechanical ventilation methods were completely unacceptable.

A natural ventilation system from SE Controls / NVS provided the answer. Combined with large areas of exposed thermal mass that reduce the need for cooling during warm periods (concrete panels to the first floor and roof soffit that give off 'coolth' to the office space during a working day), the solution works by expelling stale air and introducing fresh air through the use of climatic changes.

The natural solution

Consuming very little energy and with very little carbon emissions, natural ventilation systems demonstrate how buildings can be architecturally exciting while still being efficient, sustainable and, importantly, providing best value.

New low energy heating and mechanical ventilation systems can offer impressive energy ratings in a

similar league as natural ventilation strategies, but cost implications including capital equipment and maintenance can make them significantly more expensive.

SE Controls / NVS displayed its first-rate partnering and project management approach by working with M&E engineers Max Fordham LLP, Schüco (Glazing Systems Supplier) and Parry Bowen (Glazing Fabricator) to design, install and commission a natural ventilation system that would fit perfectly with the building's design principles.

Working with the building's sense of airiness and open plan space, the winning strategy has enabled the scheme to achieve an 'excellent' BRE Environmental Assessment Method rating. This is alongside providing occupants with the flexible, open plan and healthy working environment they desired from the outset.

An innovative ventilation strategy

Based around the company's OSO controller and a series of different actuators, the system was fitted to the façade windows and banks of roof vents. In total, 307 actuators were installed on 213 windows and 156 actuators on 78 roof vents.

The innovative OSO Controller acts as an intelligent interface between the actuators and BMS (Building Management System), with which it communicates using the LonWorks interface. As required, the BMS communicates with the OSO Control Systems and appropriately opens or closes windows and rooflights around the building, depending on the

temperature signals received from the sensors placed around the offices. Though SE Controls / NVS were able to offer a full stand alone battery backed control system, there were clear technical and financial advantages in allowing the BMS to be in overall control – a flexible approach encouraged by SE Controls / NVS to give clients a best value solution.

‘The fact that occupants could control the vents with override switches was also a very important element.’
Guy Nevill
Max Fordham LLP

As the temperature inside increases above a set parameter, the warm stale air rises. The rooflights are opened to allow this warm air to escape, rather than filling the ceiling space - and eventually the whole room - from the top down. The façade windows are opened to allow cool fresh air to be drawn in, replacing the stale air. This is a typical application of the stack ventilation principal, utilising the buoyancy effect of warm air to create low pressure areas at the bottom of the stack. This in turn draws replacement air from wherever possible – ideally, in this case, an open window. As the temperature levels return to normal, the windows and rooflights close incrementally.

During the summer, a night time cooling strategy is triggered - in part due to a comparison of internal and external temperatures during the day.

SE CONTROLS
secontrols.com

> **'Last year, temperature targets for the building were met which was great news.'**
> Guy Nevill
> Max Fordham LLP

If applicable, the strategy results in signals from the BMS to the relevant OSO controllers to open the vents, allowing cool night air to enter the building and purge the interior warm air.

The thermal mass of the building simultaneously plays its part in the cooling strategy by cooling down over night. As a result, in the morning there is a degree of stored 'coolth' in the exposed concrete structures.

Then during the day, warm air is cooled by these surfaces, which in turn warm up ready for the next night cooling cycle. Such a cooling strategy can typically reduce internal temperatures by a further 1°C during the day.

During the winter, heat produced by people and electrical equipment in the office is retained as much as possible by high performance building insulation - and by minimising the opening of windows and vents, reducing the need for mechanical heating.

When ventilation is required, it is important to minimise draughts and to maximise heat retention. This ensures that much of the heat energy carried in the air is transferred from the escaping warm stale air to the incoming cold fresh air. A reduction in the chilling effect of the incoming air is achieved by tempering it, which in turn reduces cold draughts. This is further encouraged by Max Fordham's introduction of heat exchangers mounted in a number of the roof turrets (affectionately called 'snouts'). In the winter, the heat energy in the escaping warm air is absorbed by the heat exchangers and passed via pipework to the heating trenches, where it contributes to the active heating of the building.

It has been shown that for occupants of a building to truly buy into an adaptive natural ventilation system, people need to feel they have a degree of personal control over it. This is no different with The National Trust's

employees. To provide them with such control, each of the banks of windows and rooflights has an override switch connected directly to the SE Controls OSO controllers.

This allows them to be opened or closed as required, while relaying information to the BMS so that it knows the exact position of every actuator around the building.

When a switch is pressed, the OSO controller locks out the BMS control signal it is receiving, ignoring it for a preset amount of time (typically 60 minutes).

Fig.1

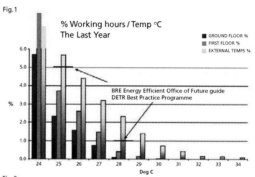

Fig.2

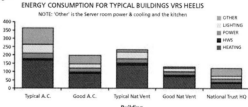

Heelis performance data and graphic Courtesy of Max Fordham LLP

At the end of this manual override time period, the OSO controller switches back to automatic, responding directly to signals from the BMS system again. However, if there is an emergency during a lockout period, the BMS can always prematurely regain control of the OSO by sending a separate reset signal.

Delivering tangible benefits

The innovative and integrated approach taken in the design of the building and its systems has resulted in the Heelis building being one of the most energy efficient, naturally ventilated offices in the UK. Staff are finding that working in an office with adaptive natural ventilation and an intelligent environmental system means fewer complaints, and no reported ill health or discomfort from air conditioned stuffiness.

Temperature levels in the office during the last year, including the unusually long and hot summer, proved to be well within recommended guidelines set by the DETR Best Practice Programme. This states that temperatures should climb to 25°C or above for no more than 5% of the working year, and to 28°C or above for no more than 1% of the working year. The graph above from Max Fordham LLP clearly shows both ground floor

secontrols.com

and first floor temperatures to easily meet these criteria.

Heelis Facilities Management team has taken a very proactive approach to ensure staff understand all the sustainable features of the building. All existing and new staff are trained on why and how the various systems operate.

It is explained that, like all buildings, there will be days when some staff will feel the environment is either too cold or too hot. As such they are encouraged to fine tune their own comfort by wearing layers. A quick look around the office after hours proves this, as almost every chair has a jacket or cardigan hanging on its backrest.

Liz Adams, Property Manager for Heelis, and her colleagues have even nicknamed the building "a cardigan building". Liz is convinced that staff education, which is ongoing and includes weekly news update bulletins, is key to the wholehearted acceptance The Trust staff have given towards their new working environment.

Along with photovoltaic solar panels and other energy saving initiatives, the natural ventilation strategy has meant that Heelis is expected to have one of the lowest carbon dioxide emissions ratings of any building in the country. The overall target for the building was ambitiously set by Max Fordham LLP at just 15kg CO_2 per square metre per year, compared with 169kg for a an inefficient air conditioned office (with a typically under-maintained air conditioning system). This target is very close to being met; Max Fordam LLP has shown that Heelis has achieved 21Kg CO_2 per m^2 per year (excluding catering and communications plant). This equates to almost a third of that achievable by a building equipped with a highly efficient air conditioning system, and only 13% of the CO_2 footprint of an inefficient air conditioned building.

The bar chart (Fig. 2 previous page) shows the energy consumption figures

for Heelis compared to other typical buildings. If the 'Other' category is ignored - with Heelis this includes the public café with its various ovens and refrigeration units, as well as significant IT server loads - one can see that Heelis performs significantly better than a 'good' nat vent building.

Initial calculations show that The National Trust's sustainable design approach to Heelis has also generated massive initial cost savings of £550,000 a year – thanks in part to choosing natural ventilation over air conditioning. The Trust also expects to save around £400,000 annually in future London weighting payments and more than £250,000 in administrative costs, thanks to the improved working practices allowed by the new building that pull together The Trust's different multi-site teams to one site at Swindon.

Award-winning praise
The multi award winning Heelis building has been recognised by RIBA for its eco-friendly attributes; RIBA presented the scheme with its prestigious Sustainability Award. Judges described the project as having:

"Sustainable design that is quite simple, but well delivered. A well handled natural ventilation system with a degree of user control, super insulation, PVs, lots of daylight and sensor controlled lighting".

"The strategy delivers an exceptionally pleasant working environment. It somehow feels healthy without being worthy. There is something very direct about the strategy that makes it understandable to the occupants and general public, which means important lessons can be passed on."

Heelis has also achieved success by scooping the British Council for Offices Innovation Award 2006. Judges said of the scheme:

"It is blessed with a most ingenious and successful strategy for naturally ventilating and cooling the building. Visited by the National Panel on what was one of the hottest days of the year, it was clear the strategy was a proven success."

"Along with natural ventilation, together with the solid south slope of the roofs covered with photovoltaic panels, helped achieve an excellent BREEAM rating. The scheme shows how older forms can be efficient and green without sacrificing value. Heelis was built to a tight budget, yet has emerged as attractive as any of the National Trust's charges."

The project has also been presented with the Brick Development Association Awards 2005 for Best Commercial Building and The International FX Interior Design Awards 2005 for Best Medium/Large Office Building.

'SE Controls was very
responsive to work with.
The natural ventilation
system works well,
as does the speed
and responsiveness of
the actuators.'
Liz Adams
National Trust Property Manager

SE Controls
Lancaster House
Wellington Crescent
Fradley Park
Lichfield
Staffordshire WS13 8RZ

Tel: +44 (0)1543 443060
Fax: +44 (0)1543 443070

Email: sales@secontrols.com
Visit us at: www.secontrols.com

SE Controls is a Registered Trademark

Chapter 5

Heat Gains in Buildings

Learning outcome

- Demonstrate the various sources of heat gain in a building — Knowledge and understanding
- Describe the criteria for comfortable light intensity in a working space — Knowledge and understanding
- Describe the different lights available and their energy and functionality performance — Analysis
- Describe the procedure for estimating energy saving for lighting — Knowledge and understanding
- Describe the method of evaluating energy gains from occupants and equipment — Analysis
- Calculate a building's energy gains — Problem solving
- Examine possible energy-efficient proposals related to optimising internal gains — Reflections
- Practise further tutorial problems — Problem solving

Energy Audits: A Workbook for Energy Management in Buildings, First Edition.
Tarik Al-Shemmeri.
© 2011 Blackwell Publishing Ltd. Published 2011 by Blackwell Publishing Ltd.

5.1 Introduction

Heating load calculations for cooler climates such as the United Kingdom are based on estimating the combined fabric and ventilation heat losses. Casual heat gains may provide a significant source of energy which can be used to offset some of the heating load. Casual heat gains come from solar energy, occupants, machines and lights.

Estimation of the heat gains from lights is an area that is considered closely in this chapter. Lighting is usually wholly electric; it constitutes about 10 to 15% of the total energy cost in residential buildings and about 25% in commercial buildings. There is a simple rule for the lighting requirement in buildings: in order to provide sufficient light, the energy consumption involved is in the region of 20-30 W/m².

The eye recognises waves in the range between the infrared and the ultraviolet parts of the electromagnetic spectrum (Figure 5.1). Colours are not received with equal sensitivity - the eye being least sensitive to red and violet and most sensitive to the yellow-green part of the spectrum.

5.2 Lighting

Photometry deals with the measurement of visible light as perceived by human eyes. The human eye can only see light in the visible spectrum and has different sensitivities to light of different wavelengths within the spectrum.

Artificial lighting constitutes a significant part of all electrical energy consumed worldwide. In homes and offices, between 20 and 50% of the total energy consumed is due to lighting. Most importantly, in some buildings, over 90% of lighting energy consumed constitutes an unnecessary expense because the building is over-illuminated.

The cost of such lighting can be substantial. A single 100 W light bulb used for just 6 hours a day can cost £21.90 per year to use (£0.10/kWh). Replacing such light bulbs with more efficient ones will save more than 50% of the cost of electricity - £10.95. In addition, and because the low-energy bulb will last around ten times longer than a standard bulb, it could save you around £100 before it needs replacing.

Imagine the unnecessary overspending caused by using multiple lights in a room. Even more wasteful is leaving lights switched on when the rooms are unoccupied.

5.2.1 Lighting criteria

Lighting is required to provide a comfortable reading and working environment. Inadequate light intensity causes discomfort and unnecessary strain, resulting in reduced work efficiency and deterioration in health.

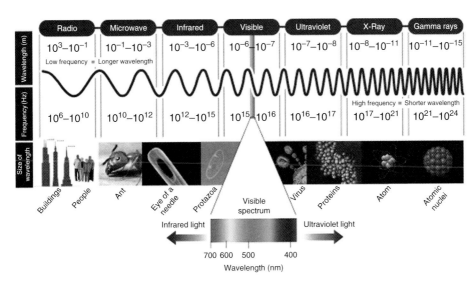

Figure 5.1   Visible light spectrum.

Many factors affect the design of a particular lighting system, such as:

● quality of life;
● type of activity;
● natural light availability;
● colour quality and glare;
● energy use;
● cost.

5.2.2   Lighting terminology

Candela (cd) is the luminous intensity, in a given direction, of a source that emits monochromatic radiation of frequency 540×10^{12} Hertz and that has a radiant intensity in that direction of 1/683 Watts per steradian.

Lux is a unit of illumination equal to the direct illumination on a surface that is everywhere one metre from a uniform point source of one candle intensity or equal to one lumen per square metre. One lux is the equivalent of 1.46 milliwatts of radiant electromagnetic (EM) power at a frequency of 540 teraHertz (540 THz or 5.40×10^{14} Hz), impinging at a right angle on a surface whose area is one square metre. A frequency of 540 THz corresponds to a wavelength of about 555 nanometres (nm), which is in the middle of the visible-light spectrum.

Illuminance is the intensity of illumination of a surface expressed in lumen/m² or Lux. The inverse law (Figure 5.2) states that the quantity of light is inversely proportional to the square of its distance:

$$\text{Illuminance}_x = \frac{\text{Illuminance at source}}{x^2} \qquad [5.1]$$

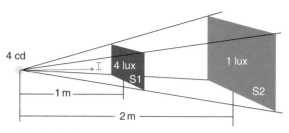

Figure 5.2 Light dissipation (inverse law).

A light is placed 1m away from the subject. If the distance is doubled to two metres, the square of its distance is 4m. The inverse of 4 is 1/4. Therefore, the quantity of light at 2m from the subject is 1/4 the amount of light at 1m. If the light is moved to a distance of 10m, the square of its distance is 100m, the inverse of which is 1/100. The quantity of light at 10m from the subject is 1% of the light at 1m from the source.

Efficacy is the total light output of a light source divided by the total power input. Efficacy is expressed in lumens per Watt:

$$\text{Efficacy} = \frac{\text{Lumens flux output}}{\text{Electrical power input}} \quad \text{units, (lumens/watt)} \qquad [5.2]$$

Since the value of this term is greater than one, the lighting industry correctly refers to it as efficacy rather than efficiency.

The CIBSE Code for Lighting recommends a maintained illuminance of 500 lux for general offices (e.g. writing, typing, reading, data processing, etc.) and for CAD work stations and conference/meeting rooms. Where the main task is less demanding, e.g. filing, a lower level of 300 lux is recommended.

5.2.3 Measurement of light intensity

A device called a photovoltaic cell, similar to the light meter used by photographers, is frequently employed in light intensity measurements. When the cell is illuminated, electrons cross the rectifier junction and produce an electromotive force (emf), which, in turn, causes current to flow if the external circuit is closed. A sensitive meter records this current, which is proportional to the light flux falling on the cell. The response of photovoltaic cells is different with differently coloured light sources and does not correspond very closely to the spectral response of the human eye. However, light filters can correct this.

5.2.4 Types of lamp

There are many types of lamp used in day-to-day lighting; see Table 5.1. They can be grouped into the following:

Table 5.1 Characteristics of electric lamps.

Lamp type (code)	Wattage range	Typical efficacy (lm/W)	Nominal life (hours)	Colour temperature (K)	Typical applications
Tungsten filament (GLS)	40–200	12	1000	2700	Homes; hotels and restaurants
Tungsten–halogen (T–H)	300–2000	21	2000–4000	2800–3000	Area and display lighting
Compact fluorescent (CFL)	9–20	60	8000	3000	Homes; offices and public buildings
Tubular fluorescent (MCF)	20–125	60	8000+	3000–6500	Offices and shops
Mercury fluorescent (MCB)	50–2000	60	8000+	4000	Factories and roadways
Mercury halide (MBI)	25–350	70	8000+	4200	Factories and shops
Low-pressure sodium (SOx)	35–180	180 (at 180W)	8000+	NA	Roadways and area lighting
High-pressure sodium (SOx)	70–1000	125 (at 400W)	8000+	2100	Factories and roadways

- the incandescent tungsten filaments lamp;
- tungsten halogen lamps;
- fluorescent lamps;
- mercury halide lamps;
- low-pressure and high-pressure sodium lamps;
- light-emitting diodes.

When electricity flows through a tube, it causes the vaporised mercury to give off ultraviolet energy. This energy then strikes phosphors that coat the inside of the lamp, giving off visible light.

Mercury vapour lamps have two bulbs – the arc tube (made of quartz) is inside a protecting glass bulb. The arc tube contains mercury vapour at a higher pressure than that of the fluorescent lamp, thus allowing the vapour lamp to produce light without using the phosphor coating.

Neon lamps are glass tubes filled with neon gas which glow when an electric discharge takes place in them. The colour of the light is determined by the gas mixture; pure neon gas gives off red light.

Metal halide lamps, used primarily outdoors for stadiums and roadways, contain chemical compounds of metal and a halogen. This type of lamp works in much the same fashion as the mercury vapour lamps except that a metal halide can produce a more natural colour balance when used without phosphors.

High-pressure sodium lamps are also similar to mercury vapour lamps; however, the arc tube is made of aluminium oxide instead of quartz, and it contains a solid mixture of sodium and mercury.

A light-emitting diode (LED) is basically a semiconductor diode. When the diode is switched on, electrons are able to recombine with electron holes, releasing energy in the form of light. The LED is usually less than 1mm squared in area and contains components to shape its reflection and radiation pattern. LEDs are economical, efficient, durable and small. They are used in home, theatre and automotive lighting, traffic signals, text and video displays and communications technology, to name but a few of the growing number of applications.

5.3   Energy-saving measures for lighting

The easiest way to control light is to turn lights on manually when needed and off before going to sleep or when leaving the room. However, for more complex situations, more sophisticated methods are needed. In such cases, and in order to promote energy efficiency, lighting control is implemented by the following methods:

- *Timer control.* Here, timers are set to switch off lighting for periods of known inactivity, such as the end of the working day.

- *Daylight control.* Lights are switched on or off, or dimmed, according to the level of daylight detected in a room.
- *Occupation control.* Sensors detect noise or movement in an area. The sensors turn lighting on when there is someone in the area and off again after a time delay if there is nobody in the room.
- *Local switching.* Here, it is only possible to switch lights on in the part of the room which is being occupied.
- *Use of energy-efficient lights.* New and more efficient light bulbs use much less energy and last longer than the old tungsten bulbs.

It is possible to ensure that the cost of energy related to lighting is kept at a minimum by regular maintenance, which involves periodic cleaning of the lighting fixtures and lamps.

5.4   Casual heat gains from appliances

Equipment and appliances can be either electric or, less frequently, battery operated. That said, some process work may derive its energy from nonelectric sources such as steam or compressed gas. At home, washing machines, radios and television sets are electric while cookers may be either electric or gas.

Electrical appliances are now given an 'energy rating' indicator. The labels displayed on machines are rated on a scale from A–G, where A represents the most efficient and G the least efficient machine.

Whatever the energy input, it is important to point out that energy is provided at a price, and the net output is proportional to the method of provision. For example, space heating can be achieved via an electric heater, which usually has an efficiency of 95% or better at a price of, say, 8 pence per kWh; the same duty can be fulfilled by an open fire using natural gas (70–80%) or liquid fuel (60–70%) or coal (60% or under). These efficiencies are typical values which may vary according to design and manufacture. The cost of natural gas is about 2.5 p/kWh; others vary quite considerably.

It is important to state that the heat released from any combustible substance depends on the following factors:

- its calorific value;
- the mass of fuel burnt in a given time;
- the efficiency of burning (this, in itself, depends on the design of the burner, the speed of combustion and the adequacy of air supply necessary for the combustion). Refer to the manufacturer's specification for the efficiency of a specific product.

The heat release rate due to burning of a fuel is given by:

$$\text{Heat release (kW)} = \text{mass (kg/s)} \times \text{calorific value (kJ/kg)} \times \text{efficiency of combustion (fraction)} \qquad [5.3]$$

Table 5.2 Typical casual heat sources.

Type of heat source	Typical heat emission
Adults	
Sleeping	80 W
Sitting quietly	120 W
Walking slowly	230 W
Medium work	265 W
Heavy work	570 W
Lights and electricals	
Fluorescent @ 400 lux	20 W/m² floor area
Tungsten @ 200 lux	40 W/m² floor area
Desktop computer	150 W
Computer printer	100 W
Visual display unit	200 W
Photocopier	800 W
Hair dryer	800 W
Domestic fridge-freezer	150 W
Colour TV	100 W
Hi-fi unit	100 W
Toaster	500 W
Oven	2500 W

5.5 Occupants' heat gains

Occupants produce heat, naturally, as by-product of metabolism. The quantity of heat produced by humans depends on several factors:

- age;
- gender;
- type of activity they are undertaking (see Table 5.2).

5.6 Worked examples

Worked example 5.1

The lighting requirements of a sports centre are being met by 200 40 W standard fluorescent lamps. The lamps are close to completing their service life and are to be replaced by their 30 W high-efficiency counterparts. The standard and high-efficiency fluorescent lamps can be purchased at quantity at a cost of £1.00 and £7.00 each, respectively. The facility operates for 2800 hours a year and all of the lamps are kept on during operating hours. Taking the unit cost of electricity to be £0.07/kWh, determine how much energy and money will be saved a year as a result of switching to the high-efficiency fluorescent lamps. Also determine the simple pay-back period assuming a fixed price for electricity.

Solution:

The annual cost of energy consumed is determined by multiplying the total number of units used (kWh) by the total number of hours (h) and the cost of each unit (£/kWh).

Consider first the 40W bulbs. The annual cost is:

$$C_1 = 200 \times \frac{40W}{1000} \times 2800 \frac{h}{y} \times 0.07 = £1568$$

Now consider the replacement 30W lights. The annual cost now is:

$$C_2 = 200 \times \frac{30W}{1000} \times 2800 \times 0.07 = £1176$$

The saving made is calculated as the difference between the costs of the old and the new: Saving/year = £1568 − £1176 = £392.

Cost difference of replacing lights = 200 × (£7 − £1) = £1200

$$\text{Pay-back} = \frac{\text{capital}}{\text{saving}/y} = \frac{£1200}{£392/y} = 3 \text{ years}$$

Worked example 5.2

A 20-year-old natural gas central heating boiler has an efficiency of 60% and is in need of replacement. Two options are available: a conventional boiler that has an efficiency of 70% and costs £900, and a high-efficiency condensing boiler that has an efficiency of 90% and costs £1500. If the heating bill is £1100 a year, which option do you recommend?

Solution:

The relevant information is tabulated in Worked example 5.2, Table 1.

Worked example 5.2, Table 1

Boiler	Capital cost	Efficiency	Operating cost	Pay-back
Current	NA	60%	£1100	NA
Conventional	£900	70%	(0.6/0.7) × 1100 = £942.85	900/ (1100–942.8) = 5.7 years
Condensing	£1500	90%	(0.6/0.9) × 1100 = £733	1500/ (1100–733) = 4.1 years

Note that the pay-back period is calculated as the simple ratio of capital cost divided by the saving made annually from the investment. Hence, although the high-efficiency boiler is more expensive to buy, it pays for itself before the conventional boiler.

Worked example 5.3

A university campus has 200 classrooms and 100 faculty offices. Each classroom is equipped with six fluorescent light bulbs, each consuming 60W. The faculty offices, on average, have half as many light bulbs. The campus is open 240 days a year, between 9am and 9pm. The classrooms and faculty offices are unoccupied for an average of 4 hours a day, but the lights are kept on. If the unit cost of electricity is £0.075/kWh, determine how much energy and money the campus will save a year if the lights in the classrooms and faculty offices are turned off during unoccupied periods. If light sensors costing £60 each are installed, what is the pay-back period?

Solution:

The annual cost of energy consumed is determined by multiplying the total number of units used (kWh) by the total number of hours (h) and the cost of each unit (£/kWh).

$$\text{Original cost} = (200 \times 6 \times 60 + 100 \times 3 \times 60) \times \frac{240}{1000} \times 12 \times 0.075$$

$$= £19\,440$$

Saving to switch off 4h/d = (4/12) × 19 440 = £6480
Cost of installing a sensor in each room = (200 + 100) × £60 = £18 000

$$\text{Pay-back} = \frac{\text{Capital}}{\text{Saving/y}}$$

$$\text{Pay-back} = \frac{18000}{6480} = 2.77 \text{ years}$$

Worked example 5.4

An exercise room has five weight-lifting machines and five treadmills, each equipped with a 3 kW motor (efficiency of utilisation 80%). During peak evening hours, all pieces of exercising equipment are used continuously, and there are also four people doing light exercises while waiting in line for one piece of the equipment. Determine the rate of heat gain of the exercise room from people and the equipment under peak load conditions. Assume the average heat loss for this type of activity is 525 W per person.

Solution:

The weight-lifting machines do not have any motors and thus they do not contribute directly to the internal heat gain. The total heat generated by the motors is proportional to the motor rating, the energy conversion efficiency and the number of machines in operation:

$$\begin{aligned} Q_{motors} &= (\text{No of motors}) \times W_{motor} \times \text{efficiency} \\ &= 5 \times 3000 \times 0.80 \\ &= 12\,000 \text{W} \end{aligned}$$

The heat gain from 14 people is:

$$\begin{aligned} Q_{people} &= (\text{No of people}) \times Q_{person} \\ &= 14 \times (525 \text{ W}) \\ &= 7350 \text{ W} \end{aligned}$$

So the total rate of heat gain is:

$$\begin{aligned} Q_{total} &= Q_{motors} + Q_{people} \\ &= 12\,000 + 7350 \\ &= 19\,350 \text{ W} \end{aligned}$$

5.7 Tutorial problems

5.1 The lighting requirements of a sports centre are being met by 200 60 W standard light bulbs. The lights are close to completing their service life and are to be replaced by 30 W high-efficiency fluorescent lamps; both can be purchased in quantity at a cost of £1.00 and £9.00 each, respectively. The facility operates for 2800 hours a year and all of the lamps are kept on during operating hours. Taking the unit cost of electricity to be 7 p/kWh, determine whether the investment in the fluorescent lights is justifiable.

Ans. (Yes, pay-back in 1.4 years)

5.2 You have been given the task of purchasing a replacement boiler for the swimming baths of a large sports complex. The present heating bill is £2000/year and the current boiler has 50% efficiency. There are two types of replacement boiler on the market, with the specifications listed in Tutorial problem 5.2, Table 1. Select the best option and justify your answer by calculations.

Tutorial problem 5.2, Table 1

Boiler type	Capital cost	Efficiency (%)
Conventional	£3600	75
Condensing	£4000	92

Ans. (Condensing)

5.3 An exercise room has ten weight-lifting machines and ten tread-mills, each equipped with a 3 kW motor (efficiency of utilisation 80%). During peak evening hours, all pieces of exercising equipment are used continuously, and there are also four people doing light exercises while waiting in line for one piece of the equipment. Determine the rate of heat gain of the exercise room from people and the equipment under peak load conditions. Assume the average heat loss for this type of activity is 525 W per person.

Ans. (36.6 kW)

5.8 Case Study: Calculation of heating load for a building – options

In this section, a typical building is considered. The heating load is calculated and this is followed by a step-by-step investigation of the few options available, the heating load is then recalculated and a table of the different solutions presented.

The building is a detached single-storey building (length 20 m, width 10 m and height 4 m). It is maintained at 20°C while the outside average temperature is 5°C. The building has a ventilation rate of 1 ACH, 1kW rating. A five-aside football game is being played inside the gym (100W/person), and there are twenty 50 W light bulbs switched on continuously.

Calculate the building heat losses and gains, and hence deduce the building's heating requirement. Relevant data are given in Case Study 5.8, Table 1, which you are required to complete.

Case Study 5.8, Table 1 Building thermal data.

Element	U value (W/m²K)
Door	3
Windows	6
Walls	3.3
Roof	3.25
Floor	1.47

5.8.1 Scenario 1: The original building description

The relevant information for this scenario is shown in Case Study 5.8, Table 1.

Case Study 5.8, Table 2

Element	U value (W/m²K)	Area (m²)	Temperature difference (°C)	Heat loss (W)
Door	3	4	15	180
Windows	6	20 total	15	1800
Walls	3.3	216	15	10 692
Roof	3.25	200	15	9750
Floor	1.47	200	15	4410
Fabric heat loss = Total (Doors, Windows, Walls, Roof and Floor) =				−26 832
Ventilation heat loss =		0.335 N V ΔT		−4020
Heat gains from occupants =		10×100		+1000
Heat gains from lights =		20×50		+1000
Heat gains from machines =		1×1000		+1000
Net heat transfer for the building =				−27 852

5.8.2 Scenario 2: The addition of loft insulation

The relevant information for this scenario is shown in Case Study 5.8, Table 3. (Loft insulation $U = 0.45$.). All data as in Scenario 1, except for the roof.

Case Study 5.8, Table 3.

Element	U value (W/m²K)	Area (m²)	Temperature difference (°C)	Heat loss (W)
Door	3	4	15	180
Windows	6	20 total	15	1800
Walls	3.30	216	15	10 692
Roof	0.45	200	15	1350
Floor	1.47	200	15	4410
Fabric heat loss = Total (Doors, Windows, Walls, Roof and Floor) =				−18 432
Ventilation heat loss =			0.335 N V ΔT	−4020
Heat gains from occupants =			10 × 100	+1000
Heat gains from lights =			20 × 50	+1000
Heat gains from machines =			1 × 1000	+1000
Net heat transfer for the building =				−19 452

5.8.3 Scenario 3: The addition of cavity insulation

The relevant information for this scenario is shown in Case Study 5.8, Table 4. (Cavity insulation $U = 0.45$.). All data as in Scenario 1, except for the walls.

Case Study 5.8, Table 4

Element	U value (W/m²K)	Area (m²)	Temperature difference (°C)	Heat loss (W)
Door	3	4	15	180
Windows	6	20 total	15	1800
Walls	0.45	216	15	1458
Roof	3.25	200	15	9750
Floor	1.47	200	15	4410
Fabric heat loss = Total (Doors, Windows, Walls, Roof and Floor) =				−17 598
Ventilation heat loss =			0.335 N V ΔT	−4020
Heat gains from occupants =				+1000
Heat gains from lights =				+1000
Heat gains from machines =				+1000
Net heat transfer for the building =				−18 618

5.8.4   Scenario 4: Windows upgrade to double glazing

The relevant information for this scenario is shown in Case Study 5.8, Table 5. (Double glazed windows *U* value = 3.). All data as in Scenario 1, except for the windows.

Case Study 5.8, Table 5

Element	U value (W/m²K)	Area (m²)	Temperature difference (°C)	Heat loss (W)
Door	3	4	15	180
Windows	3	20 total	15	900
Walls	3.30	216	15	10 692
Roof	3.25	200	15	9750
Floor	1.47	200	15	4410
Fabric heat loss = Total (Doors, Windows, Walls, Roof and Floor) =				−25 932
Ventilation heat loss =		0.335 N V ΔT		−4020
Heat gains from occupants =				+1000
Heat gains from lights =				+1000
Heat gains from machines =				+1000
Net heat transfer for the building =				−26 952

5.8.5   Scenario 5: The introduction of all three upgrades

The combined information for this scenario is shown in Case Study 5.8, Table 6.

Case Study 5.8, Table 6

Element	U value (W/m²K)	Area (m²)	Temperature difference (°C)	Heat loss (W)
Door	3	4	15	180
Windows	3	20 total	15	900
Walls	0.45	216	15	1458
Roof	0.45	200	15	1350
Floor	1.47	200	15	4410
Fabric heat loss = Total (Doors, Windows, Walls, Roof and Floor) =				−8298
Ventilation heat loss =		0.335 N V ΔT		−4020
Heat gains from occupants =				+1000
Heat gains from lights =				+1000
Heat gains from machines =				+1000
Net heat transfer for the building =				−9318

5.8.6 Results summary table

A summary of the above results is shown in Case Study 5.8, Table 7.

Case Study 5.8, Table 7

Upgrade	Heating load	Energy reduction (%)
1. Original	−27 852	NA
2. Loft insulation $U = 0.45$	−19 452	30
3. Cavity insulation $U = 0.45$	−18 618	33
4. Double glazing $U = 3.0$	−26 952	3
5. All upgrades	−9318	66

CONCLUSION

There are many options to consider in order to reduce energy consumption. When the costs of installation are considered, these figures may not look so exciting, *but* environmental benefits have not been taken into account in the above. Imagine the benefits from that.

Chapter 6

Thermal Comfort

Learning outcomes

• Describe the concept of thermal comfort	Knowledge and understanding
• Describe the law of conservation of energy applied to the human body	Analysis
• Distinguish between sensible and latent heat release from the human body during different activities	Knowledge and understanding
• Estimate thermal comfort condition, using the predicted mean vote and percentage of dissatisfied	Analysis
• Solve problems associated with energy conversion	Problem solving
• Practice further tutorial problems	Problem solving

Energy Audits: A Workbook for Energy Management in Buildings, First Edition.
Tarik Al-Shemmeri.
© 2011 Blackwell Publishing Ltd. Published 2011 by Blackwell Publishing Ltd.

6.1 Thermal comfort in human beings

Thermal comfort is defined as that state of mind in which satisfaction is expressed with the thermal environment. Maintaining a building environment at a higher temperature than that set out by the comfort criterion is wasteful of energy and detrimental to the efficiency of occupants.

The mechanism of thermal comfort for a human is governed by the basic modes of heat and mass transfer. Using these basic modes it is possible to establish a heat balance equation for a given person in a given activity, clothing and environment. However, the heat balance in itself is not sufficient to establish thermal comfort. In the wide range of environmental conditions where a heat balance may be achieved, there is only a narrow range for mean skin temperature and sweat loss where thermal comfort is established. The mean skin temperature and sweat loss will vary with activity for individuals, and is generally coupled with six external parameters:

- air temperature, t_{db};
- mean radiant temperature, t_r;
- relative air velocity, v_a;
- humidity, Φ;
- activity level, M;
- clothing thermal resistance, I_{cl}.

Any combination of these six parameters satisfies a *comfort condition*, giving the majority of individuals a thermally neutral sensation.

6.2 Energy balance of the human body

The thermoregulatory system of the human body maintains a constant temperature and there is, therefore, no net heat storage. The chemical energy supplied to the body may be equated to the heat and work output, as shown in Figure 6.1. The metabolic rate, M, is the total energy released by the oxidation processes in the human body per unit time; this represents the energy intake to the body. The total heat flow rate Q_T dissipated by the body may be considered to be the sum of the latent heat flow rate Q_L due to mass exchange effects such as sweating and respiration, and the sensible heat flow rate Q_S due to heat transfer from the skin, including that through any clothing to the surroundings. Another energy element involved in the balance is the work output, W. An energy balance on the human body then yields:

Energy intake (M) = energy expenditure (heat, Q_T and work, W)

or, in symbols:

$$M - W - Q_T = 0 \qquad\qquad\qquad [6.1]$$

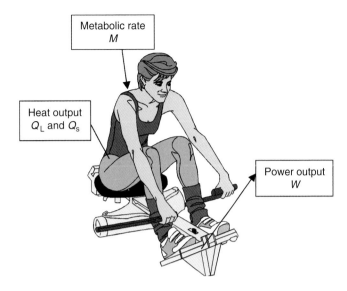

Figure 6.1 Energy balance for the human body.

Energy is released in the body by oxidation at a rate equivalent to the needs of bodily functioning. Even when completely at rest, people continue to metabolise food just to keep their organs functioning. The metabolic rate (M) varies between individuals; as a result, some have a slightly higher core body temperature than others. Generally, older people and children have slower metabolic rates (75%) while women's metabolic rate is 85% of that for men. For any person, the metabolic rate varies on a regular daily cycle. Finally, the amount of physical exercise done (W) affects the metabolic rate, since exercise requires a greater consumption of energy; this, in turn, produces more heat.

The efficiency, μ, of human activity is defined as the ratio of physical work to the metabolic rate, W/M (which is usually below 20%, and is zero when motionless).

Expressing Equation [6.1] in terms of body surface area and rearranging gives:

$$M - W = A\,(q_s + q_L) \tag{6.2}$$

where q_L is the 'latent or wet' heat flow rate per unit area of skin due to sweating, evaporation and respiratory effects, q_s is the 'sensible or dry' heat flow rate per unit area of skin due to convection and radiation. Latent and sensible heat losses for different human activities are shown in Table 6.1.

6.3 Latent heat losses

The latent heat transfer, which is accompanied by mass exchange effects, may be subdivided into three parts:

Table 6.1 Metabolic rates for different activity levels.
Courtesy of http://www.engineeringtoolbox.com/

Activity	Human metabolic rate	
	(W/m²)	(Met)
Lying down	46	0.8
Sitting, relaxed	58	1.0
Sitting activity (office work, school, etc.)	70	1.2
Standing activity (shop, laboratory, etc.)	93	1.6
Moving activity (house work, working at machines, etc.)	116	2.0
Harder activity (hard work at machines, workshops, etc.)	165	2.8

- the heat flux due to water diffusion through the skin to the surroundings, q_d;
- the heat flux due to evaporation of sweat on the skin surface, q_e;
- the respiration heat loss due to exhaling warm/inhaling cooler air, q_{res}.

i.e.

$$q_L = q_d + q_e + q_{res}$$ [6.3]

6.3.1 Heat loss by diffusion

The amount of heat loss due to water diffusion takes place all the time and is not controlled by the thermoregulatory system, it can be calculated by:

$$q_d = 3.05 \times 10^{-3} \, (256 \, t_s - 3373 - P_s)$$ [6.4]

Equation [6.4] indicates that heat transfer by diffusion is strongly dependent on two variables: the skin temperature, which can be approximated by:

$$t_s = 35.7 - 0.0275 \, (M - W)$$ [6.5]

and the water vapour pressure, P_s, which is a function of the relative humidity and temperature of the ambient air, and is given by:

$$P_s = \phi \times 0.6105 \times e^{\frac{17.27 \times t_{db}}{237.3 + t_{db}}}$$ [6.6]

with Φ being the relative humidity (fraction) and t_{db} the dry-bulb temperature (°C).

For example, for an average metabolic rate of 1.2 met, an ambient air temperature of 23°C and a relative humidity of 50%, the heat dissipation by diffusion is approximately $q_d = 12 \, W/m^2$.

6.3.2 Heat loss by evaporation

The amount of heat loss due to sweat is closely related to skin temperature. It is one of the most effective ways by which the body maintains its core temperature and avoids overheating during heavy physical activity. The amount of heat loss via sweat is given by:

$$q_e = 0.42 \, (M - W - 58.15) \qquad [6.7]$$

6.3.3 Heat loss by respiration

The air necessary for metabolism is taken from the surroundings, which are normally at a temperature lower than the body temperature of 37°C. The temperature of exhaled air is in the region of 34°C. This heat exchange is due to the difference in temperature of the air leaving the lungs and that entering, and since this air is moist, the equation to calculate the heat loss is made up of two terms - sensible and latent - shown in the following expression:

$$q_{res} = 0.0014 \, M \, (34 - t_{db}) + 1.72 \times 10^{-5} \, M \, (5867 - P_s) \qquad [6.8]$$

For typical indoor activity and an ambient temperature of 20°C, the heat loss by respiration is of the order of $5 \, W/m^2$ maximum, and hence can be neglected.

6.4 Sensible heat losses

The term q_s in Equation [6.2] refers to the heat flux through the clothing by conduction and then to the surroundings from the outer surface of clothing by convection and radiation (as shown in Figure 6.2).

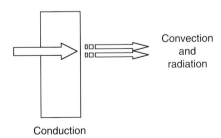

Conduction

Convection and radiation

Figure 6.2 Heat flow across the skin/clothes.

6.4.1 Heat loss by conduction

Conduction through the clothing is expressed as:

$$q_S = \frac{A_c}{A}\frac{k}{x}(t_s - t_{cl})$$ [6.9]

where the ratio A_c/A is termed the area factor for the clothes, f_{cl}, and is used to correct the equation for the fact that q_h is based on the area, A, of the body rather than the clothes. The term $Ax/A_c k$ is the total thermal resistance of the clothing and has units of m²K/kW. Thus, it is usually expressed as a 'clo' index, I_{cl}, which is defined such that for a typical summer lightweight business suit it has a value of 1 (1 clo = 0.155 m²K/W). Various values of clo for different garments are shown in Table 6.2. Hence the heat transfer by conduction can be expressed as:

$$q_S = \frac{(t_s - t_{cl})}{0.155 \times I_{cl}}$$ [6.10]

The clothes temperature, t_{cl}, is calculated by the following expression:

$$t_{cl} = 35.7 - 0.028\,(M - W) - 0.155\,I_{cl}\,(M - W) - 0.00305\,(5733 - 6.99$$
$$(M - W) - P_s) - 0.42\,((M - W) - 58.15) - 0.000017M\,(5867 - P_s)$$
$$- 0.0014M\,(34 - t_{db})$$ [6.11]

6.4.2 Heat loss by convection

Convection from the bare skin or through the clothes is given by:

$$q_c = h_c\,f_{cl}\,(t_{cl} - t_{db})$$ [6.12]

where the clothing factor is expressed as:

$$f_{cl} = 1.0 + 0.2\,I_{cl} \qquad \text{for } I_{cl} < 0.5 \text{ clo}$$ [6.13]

$$= 1.05 + 0.1\,I_{cl} \qquad \text{for } I_{cl} > 0.5 \text{ clo}$$ [6.14]

The convection heat transfer coefficient is expressed as:

$$h_c = 2.38\,(t_{cl} - t_{db})^{0.25} \quad \text{free convection (no air movement)}$$ [6.15]

$$h_c = 12.1\,(v_a)^{0.5} \qquad \text{forced convection (with air movement)}$$ [6.16]

where v_a is the relative velocity of air in m/s.

6.4.3 Heat loss by radiation

Radiation from the body to the surroundings is given by:

$$q_r = h_r\,f_{cl}\,(t_{cl} - t_{db})$$ [6.17]

Table 6.2   Clothing factors. Adapted from
http://www.engineeringtoolbox.com/

	Insulation	
Item of clothing	Clo	m²K/W
Panties	0.03	0.005
Briefs	0.04	0.006
Pants, long legs	0.1	0.016
Shirt, sleeveless	0.06	0.009
T-shirt	0.09	0.014
Shirt with long sleeves	0.12	0.019
Half-slip in nylon	0.14	0.022
Short sleeve	0.09	0.029
Light blouse with long sleeves	0.15	0.023
Light shirt with long sleeves	0.2	0.031
Normal with long sleeves	0.25	0.039
Blouse with long sleeves	0.34	0.053
Shorts	0.06	0.009
Light trousers	0.2	0.031
Normal trousers	0.25	0.039
Overalls	0.28	0.043
Sleeveless vest	0.12	0.019
Thin sweater	0.2	0.031
Thick sweater	0.35	0.054
Vest	0.13	0.02
Light summer jacket	0.25	0.039
Smock	0.3	0.047
Jacket	0.35	0.054
Overalls, multi-component	0.52	0.081
Coat	0.6	0.093
Socks	0.02	0.003
Thin-soled shoes	0.02	0.003
Thick-soled shoes	0.04	0.006
Thick ankle socks	0.05	0.008
Thick long socks	0.1	0.016
Winter dress, long sleeves	0.4	0.062
Short gown, thin strap	0.15	0.023
Long gown, long sleeve	0.3	0.047
Long pyjamas with long sleeve	0.5	0.078

where the radiation constant $h_r = 3.90$ W/m²K.

Typical values of body heat balance under various conditions are shown in
Table 6.3 (heat flow rates are expressed in W/m² of skin surface and should
therefore be multiplied by the mean body area of 1.77 m² to obtain typical heat
flow rates for the total body).

Table 6.3 Sensible and latent heat losses for different human activities. Courtesy of http://www.engineeringtoolbox.com/

Human activity	Clothing (Clo)	Comfort temperature (°C)	Relative air speed (m/s)	Heat transferred from person to surroundings			
				Convection (Watts)	Radiation (Watts)	Latent vapour (Watts)	Total (Watts)
Sitting still	Naked	28.8	< 0.1	36	38	27	102
		30.1	0.3	47	29	27	102
		30.7	0.5	51	24	27	102
		31.4	1.0	57	20	27	102
	0.5	26.2	< 0.1	36	37	28	102
		27.4	0.3	47	28	28	102
		27.9	0.5	50	23	28	102
		28.5	1.0	56	19	28	102
	1.0	23.3	< 0.1	36	35	30	102
		24.5	0.3	45	27	30	102
		25.0	0.5	50	22	30	102
		25.6	1.0	55	17	30	102
	1.5	20.7	< 0.1	36	34	31	102
		21.8	0.3	45	26	31	102
		22.3	0.5	50	21	31	102
		22.8	1.0	55	16	31	102

Medium activity	Naked	24.4	< 0.1	59	65	77	204
		26.2	0.3	76	51	77	204
		27.1	0.5	83	44	77	204
		28.2	1.0	93	35	77	204
	0.5	19.9	< 0.1	60	63	80	204
		21.6	0.3	76	48	80	204
		22.4	0.5	83	41	80	204
		23.3	1.0	92	33	80	204
	1.0	15.3	< 0.1	60	59	83	204
		16.9	0.3	76	45	83	204
		17.7	0.5	83	38	83	204
		18.6	1.0	91	30	83	204
	1.5	10.9	< 0.1	62	57	84	204
		12.5	0.3	77	43	84	204
		13.2	0.5	83	36	84	204
		14.0	1.0	91	29	84	204
High activity	Naked	22.1	0.3	107	67	129	306
		23.4	0.5	117	60	129	306
		24.9	1.0	130	48	129	306
	0.5	15.7	0.3	108	64	133	306
		16.8	0.5	119	55	133	306
		18.2	1.0	130	44	133	306
	1.0	9.3	0.3	110	59	135	306
		10.4	0.5	120	51	135	306
		11.7	1.0	131	40	135	306
	1.5	3.2	0.3	113	56	137	306
		4.2	0.5	122	47	137	306
		5.4	1.0	131	37	137	306

6.5 Estimation of thermal comfort

Thermal comfort involves creating a condition such that the maximum number of people in that environment would recommend no adjustment to the temperature or relative humidity. Since it is impossible to please everybody all the time, an index has been developed to express an overall feeling of comfort in that environment.

The *PMV (Predicted Mean Vote) Index* gives a subjective thermal reaction for a large group of subjects according to the psychophysical scale. It predicts the mean value of the thermal votes of a large group of people exposed to the same environment. The PMV value is specified for seven distinct sensations: Hot (+3), Warm (+2), Slightly warm (+1), Neutral (0), Slightly cool (–1), Cool (–2) and Cold (–3).

Individual votes on the PMV Index show scatter and it is preferable, in practice, to predict the number of people likely to feel uncomfortably warm or cool, since these are the people most likely to complain about the environment. The *PPD (Predicted Percentage of Dissatisfied) Index* gives a quantitative prediction of the number of thermally dissatisfied people. The relationship between the PMV and PPD is shown in Figure 6.3. Note that even when the PMV is zero, 5% of people will be dissatisfied!

The PMV upper and lower limits are set at: –0.5 < PMV < 0.5, corresponding to a PPD < 10%, which represents acceptable indoor conditions.

6.5.1 Determination of comfort temperature, PMV and PPD

The predicted mean vote is calculated by:

$$PMV = (0.303e^{-0.036M} + 0.028)\{(M - W) - (q_c + q_r) - (q_d + q_e) - q_{res}\} \quad [6.18]$$

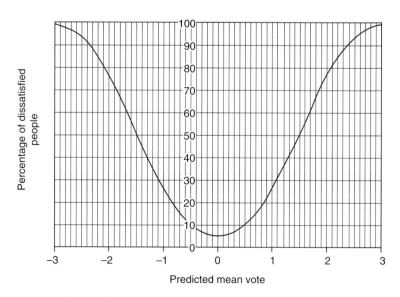

Figure 6.3 PMV–PPD relationship.

PMV typically takes values between −3 and +3; a range of accurate recommended representations is as follows:

Metabolic rate	M	46-232 W/m²
Clothing factor	I_{cl}	0-0.310 m²°C/W
Dry-bulb temperature	t_{db}	10-30°C
Radiant temperature	t_r	10-40°C
Air velocity	v_a	0-1 m/s
Vapour pressure	P_s	0-2700 Pa

The percentage of dissatisfied people is given by:

$$PPD = 100 - 95 \times e^{-(0.03353 \times PMV^4 + 0.2179 \times PMV^2)} \qquad [6.19]$$

The comfort temperature, t_{com}, is calculated as follows:

$$t_{com} = 33.5 - 3\,I_{cl} - (M/A)[0.08 + 0.05\,I_{cl}] \qquad [6.20]$$

where I_{cl} is the clothing factor in units of clo.

6.6 Worked examples

Worked example 6.1

A man walks on level ground wearing a light tracksuit (specific metabolic rate, $M = 116$ W/m² and body efficiency, $\eta = 10\%$). Assuming that the body surface area, $A = 1.77$ m²

(a) Determine the total metabolic rate, the heat loss and the muscle power output.
(b) Compare these results with those determined for his state when he returns home and is watching the TV (specific metabolic rate, $M = 58$ W/m² and body efficiency, $\eta = 5\%$). Justify any assumption required.

Solution:

(a) The total metabolic rate for this person is determined by multiplying the specific metabolic rate by the body surface area (A)

$$M = 116 \times 1.77 = 205.32 \text{ Watts}$$

By definition, the efficiency of the body, $\eta = W/M$.
Hence, the rate of work is found as:

$$W = \eta \times M = 0.10 \times 205.32 = 20.53 \text{ Watts}$$

The total heat loss $Q = M - W = 205.32 - 20.53 = 184.8$ Watts

(b) For reclining activity, a similar procedure to that used in (a) is used. Hence:

$M = 58 \times 1.77 = 102.66\,\text{W}$ (50% of the value calculated in (a))

For this case, efficiency = 5%, hence:

$W = 0.05 \times 102.66 = 5.13\,\text{W}$ (25% of the value calculated in (a))
$Q = M - W = 97.53\,\text{W}$ (53% of the value calculated in (a))

Worked example 6.2

Determine the latent heat losses by category for a person engaged in a sedentary activity ($M = 69.6\,\text{W/m}^2$) in an environment where the air temperature is 23°C and the relative humidity is 50%. Assume a skin temperature of 34°C and a physical efficiency of 10%. Assume that the body surface area, $A = 1.77\,\text{m}^2$.

Solution:

The partial pressure of water vapour is determined first:

$$P_s = \phi \times 0.6105 \times e^{\frac{17.27 \times t_{db}}{237.3 + t_{db}}} = 0.5 \times 0.6105 \times e^{\frac{17.27 \times 23}{237.3 + 23}} = 1404\,\text{Pa}$$

Heat loss by diffusion

$$\begin{aligned} q_d &= 3.05 \times 10^{-3} \times (256\,t_s - 3373 - P_s) \\ &= 3.05 \times 10^{-3} \times (256 \times 34 - 3373 - 1404) \\ &= 11.98\,\text{W/m}^2 \end{aligned}$$

Heat loss by evaporation

$$\begin{aligned} q_e &= 0.42\,(M - W - 58.15) \\ &= 0.42\,(69.6 - 0.1 \times 69.6 - 58.15) \\ &= 1.886\,\text{W/m}^2 \end{aligned}$$

Heat loss by respiration

$$\begin{aligned} q_{res} &= 0.0014\,M\,(34 - t_{db}) + 1.72 \times 10^{-5} \times M\,(5867 - P_s) \\ &= 0.0014 \times 69.6\,(34 - 23) + 1.72 \times 10^{-5} \times 69.6\,(5867 - 1404) \\ &= 6.220\,\text{W/m}^2 \end{aligned}$$

Hence, the total latent heat is the sum of all three parts found above, multiplied by the body surface area:

$$Q_L = 1.77\,[11.98 + 1.886 + 6.220] = 35.55\,\text{W}$$

Worked example 6.3

Determine the heat released by convection from a human body for two situations:

(a) When jogging at an average speed of 2 m/s.
(b) When sitting in a draught-free room.

Assume a clothing factor of 0.5 clo, a body (clothed) temperature of 34°C and an air temperature of 23°C for both cases. Assume the surface area of the human body, $A = 1.77\,m^2$.

Solution:

(a) When jogging at 2m/s, the convection coefficient

$$h_c = 12.1\, v^{0.5}$$
$$= 12.1\,(2.0)^{0.5}$$
$$= 17.1\,W/m^2K$$

The clothing factor

$$f_{cl} = 1 + 0.2\, I_{cl}$$
$$= 1 + 0.2 \times 0.5 = 1.1$$

The convection heat loss

$$Q_c = h_c\, f_{cl}\, A\, (t_{cl} - t_a)$$
$$= 17.1 \times 1.1 \times 1.77 \times (34-24)$$
$$= 366\ Watts$$

(b) For free convection, i.e. when the air velocity is zero, the convection coefficient

$$h_c = 2.38\, (t_{cl} - t_a)^{0.25}$$
$$= 2.38\,(34 - 23)^{0.25}$$
$$= 4.33\,W/m^2K$$

The convection heat loss

$$Q_c = h_c\, f_{cl}\, A\, (t_{cl} - t_a)$$
$$= 4.33 \times 1.1 \times 1.77 \times (34 - 23)$$
$$= 93\ Watts$$

Worked example 6.4

Determine the radiant heat transfer from the human body in an air-conditioned space for two situations:

(a) Clothing factor = 0.5 clo, body (clothed) temperature is 34°C and the dry-bulb temperature of the air is kept constant at 23°C.
(b) Clothing factor = 1.0 clo, body (clothed) temperature is 34°C and the dry-bulb temperature of the air is kept constant at 23°C.

Assume that the surface area of an average person's body, $A = 1.77 \, m^2$.

Solution:

(a) Since $I_{cl} = 0.5$ clo, the clothing factor, $f_{cl} = 1 + 0.2 \, I_{cl} = 1.1$.
The radiant heat loss

$$Q_r = h_r \, f_{cl} \, A \, (t_{cl} - t_a)$$
$$= 3.90 \times 1.1 \times 1.77 \times (34 - 23)$$
$$= 83.5 \text{Watts}$$

(b) Since $I_{cl} = 1.0$ clo, the clothing factor $f_{cl} = 1.05 + 0.1 \, I_{cl} = 1.15$.
The radiant heat loss

$$Q_r = h_r \, f_{cl} \, A \, (t_{cl} - t_a)$$
$$= 3.90 \times 1.15 \times 1.77 \times (34 - 23)$$
$$= 87.3 \text{ Watts}$$

Worked example 6.5

Determine the comfort temperature for personnel in a factory (light activity level 1.2 met) under the following conditions:

(a) In winter conditions with clo = 1.5.
(b) In the summer with clo = 0.5.

Solution:

The comfort temperature, t_{com}, is:

$$t_{com} = 33.5 - 3 \, I_{cl} - (M/A)[0.08 + 0.05 \, I_{cl}]$$

(a) For winter time, the comfort temperature is determined by:

$$t_{com} = 33.5 - 3 \times 1.5 - 69.6[0.08 + 0.05 \times 1.5]$$
$$= 18.21°C$$

(b) For summer time, the comfort temperature is determined by:

$$t_{com} = 33.5 - 3 \times 0.5 - 69.6[0.08 + 0.05 \times 0.5]$$
$$= 24.69°C$$

Worked example 6.6

Investigate the rate of heat dissipation by *respiration* for people when walking gently (metabolic rate per unit body area of 69.6W/m²) given a changing ambient dry-bulb temperature, t_{db}, which is varying between 5 and 40°C. Assume a relative humidity of 50% and barometric pressure of 1 bar.

Solution:

The partial pressure of water vapour is determined first:

$$P_s = \phi \times 0.6105 \times e^{\frac{17.27 \times t_{db}}{237.3 + t_{db}}} = 0.5 \times 0.6105 \times e^{\frac{17.27 \times 5}{237.3 + 5}} = 436\,\text{Pa}$$

The heat loss by respiration is:

$$q_{res} = 0.0014\,M\,(34 - t_{db}) + 1.72 \times 10^{-5} \times M\,(5867 - P_s)$$
$$q_{res} = 0.0014 \times 69.6(34 - 5) + 1.72 \times 10^{-5} \times 69.6\,(5867 - 436)$$
$$= 2.826 + 6.501 = 9.327\,\text{W/m}^2$$

The total heat loss by respiration is:

$$Q_{res} = 1.77 \times 9.327 = 16.51\,\text{W}$$

We now repeat the calculations for $t_{db} = 10\text{-}40°C$. The results are shown in Worked example 6.6, Table 1.

Worked example 6.6, Table 1

Dry-bulb temperature t_{db} °C	Vapour pressure P_s kPa	Respiration heat loss Q_{res} W
5	436	16.51
10	614	15.3
15	852	13.9
20	1168	12.4
25	1583	10.6
30	2120	8.6
35	2810	6.3
40	3686	3.6

Worked example 6.7

Investigate the rate of heat dissipation by *convection* for a female cyclist with a speed varying between 0 and 5 m/s if the air temperature, t_a, is constant at 24°C. Assume an average clothed temperature of 34°C and a clothing factor of 0.5 clo.

Solution:

For free convection ($v_a = 0$ m/s), the convection coefficient

$$h_c = 2.38 \, (t_{cl} - t_a)^{0.25}$$
$$= 2.38 \, (34 - 24)^{0.25}$$
$$= 4.23 \, W/m^2K$$

For other speeds, $h_c = 12.1 \, v_a^{0.5}$

$$f_{cl} = 1 + 0.2 \, I_{cl} = 1 + 0.2 \times 0.5 = 1.1$$
$$Q_c = h_c \, f_{cl} \, A \, (t_{cl} - t_a)$$

Worked example 6.7, Table 1 shows the results.

Worked example 6.7, Table 1

Speed (m/s)'	Convection coefficient h_{conv}	Convection heat loss Q_{con} (W)
0	4.23	82
1	12.1	235
2	17.11	333
3	20.96	408
4	24.20	471
5	27.06	527

Worked example 6.8

Investigate the rate of heat dissipation by *radiation* for an athlete running in the sunshine with a clothing factor varying between 0 and 1.5 clo if the air temperature, t_a, is constant at 14°C. Assume an average clothed body temperature of 34°C.

Solution:

The clothing factor:

$$f_{cl} = 1 + 0.2 \, I_{cl} \text{ for } I_{cl} < 1$$
$$= 1.05 + 0.1 \, I_{cl} \text{ for } I_{cl} \geq 1$$

The radiant heat coefficient, $h_r = 3.9 \, W/m^2K$ and $A = 1.77 \, m^2$.

The radiant heat loss is found using $Q_r = h_r f_{cl} A (t_{cl} - t_a)$. The results are tabulated in Worked example 6.8, Table 1.

Worked example 6.8, Table 1

Clothes thermal resistance Clo	Clothes factor f_{cl}	Radiation heat loss (Watts) Q_r
0	1	138
0.25	1.05	145
0.50	1.1	151
0.75	1.15	159
1.00	1.15	159
1.25	1.175	162
1.50	1.20	166

6.7 Tutorial problems

6.1 A man working in a factory (heavy machine work, 2.4 met) wears a light working suit (0.75 clo). Assuming a body work efficiency, $\eta = 15\%$, determine the heat loss and the muscle power output (W). Compare these results with those determined for his state when he returns home and is reclining on his sunbed (0.8 met, efficiency $\eta = 3\%$). Justify any assumption required.
Ans. (37 W, 211 W; 2 W, 79 W)

6.2 Determine the latent heat losses by category for a person engaged in heavy machining ($M = 139.2$ W/m²) in an environment where the air temperature is 23°C and the relative humidity is 50%. Assume a skin temperature of 34°C and a physical efficiency of 10%.
Ans. (19.9 W, 28.2 W, 12.8 W)

6.3 Determine the heat released by convection from a human body for two situations:

(a) When the average air velocity is 0.5 m/s.
(b) In a draught-free room.

Assume a clothing factor of 1.0 clo, a body (clothed) temperature of 34°C and an air temperature of 24°C.
Ans. (182 W, 90 W)

6.4 Determine the radiant heat transfer from the human body for two situations:

(a) Clothing factor = 0 clo, body temperature = 36°C and the dry-bulb temperature of the air = 24°C.

(b) Clothing factor = 1.0 clo, body (clothed) temperature = 32°C and the dry-bulb temperature of the air = 24°C.

Ans. (82.8 W, 63.5 W)

6.5 Determine the comfort temperature for personnel in a factory (medium activity level 1.6 met) under the following conditions:

(a) In winter conditions with clo = 1.5.

(b) In summer time with clo = 0.5.

Ans. (14.6°C, 22.2°C)

6.6 Investigate the rate of heat dissipation by respiration for a person walking gently (metabolic rate per unit body area of 69.6 W/m^2) given a changing ambient dry-bulb temperature, t_{db}, which is varying between 5 and 30°C. Assume relative humidity of 50% and a baro-metric pressure of 1 bar.
Ans. (Complete in tabular form)

6.7 Investigate the rate of heat dissipation by convection for a male runner running at an average speed of 4 m/s if the air temperature varies between 10 and 30°C. Assume an average clothed tempera-ture of 34°C and a clothing factor of 0.5 clo.
Ans. (@10°C, 1130 W)

6.8 Investigate the rate of heat dissipation by radiation for a person sunbathing in the nude with an air temperature varying between 20 and 40°C. Assume an average body temperature of 35°C, and that only half the body area (0.9 m^2) is exposed to radiative heat loss.
Ans. (@20°C, 52.65 W)

Chapter 7

Refrigeration, Heat Pumps and the Environment

Learning outcomes

- Demonstrate the need for cooling and refrigeration for buildings — Knowledge and understanding
- Reflect on the evolution of refrigeration through history and demonstrate the process of selection of refrigerants — Reflections
- Distinguish between a refrigerator and a heat pump — Analysis
- Describe the refrigeration process, cycle and components — Knowledge and understanding
- Evaluate the refrigerant properties for a typical refrigeration/heat pump cycle — Analysis
- Describe irreversibilities in the refrigeration cycle — Analysis
- Solve problems associated with refrigeration/heat pumps — Problem solving
- Practise further tutorial problems — Problem solving

Energy Audits: A Workbook for Energy Management in Buildings, First Edition.
Tarik Al-Shemmeri.
© 2011 Blackwell Publishing Ltd. Published 2011 by Blackwell Publishing Ltd.

7.1 Introduction

Refrigeration is a process of controlled removal of heat from a substance to keep it at a temperature below the ambient, and often below the freezing point of water (0°C). In principle, refrigeration is based on the fact that when a liquid is allowed to evaporate, it will absorb heat from its surroundings.

A practical method of cooling is often seen in tropical regions. People in very hot climates keep water inside a porous earthenware container; as some of this water penetrates the wall of the vessel, it evaporates on the outside surface, absorbing its latent heat from the surface and hence keeping it, and the water inside, cooler. This method of cooling, although very useful, has limitations in that the minimum temperature it can cool to is still a little below that of the ambient, and only small quantities of water can be cooled in this way. However, this method of cooling is totally environmentally friendly and is to be encouraged whenever practical.

At this stage, it is worth pointing out that there are various methods of producing a low temperature cooling environment, they are:

- *The vapour compression cycle*, in which a refrigerant undergoes a continuous change of phase (liquid/vapour) by a compression/expansion process, thus giving/taking heat away from the environment.
- *The absorption cycle*, in which a solution of varied ratio performs the cooling process.
- *The air cycle*, in which air is allowed to expand, resulting in a drop of its temperature; the extent of expansion decides the final temperature of the air.

In this text, the first type of refrigeration cycle will be considered for two reasons:

- Because at present it represents over 90% of systems in use.
- Due to the effect of common refrigerants on the environment, and to demonstrate that there are new and old refrigerants that are environmentally friendly, yet can perform the required duty.

Today, refrigeration has become an integral part of living standards; it would be unthinkable to maintain the world's food supply without refrigerated transport. On average, refrigeration represents 10 to 25% of the total energy consumption in developed countries. In the UK, for example, it is estimated that 40 000 GWh of the national annual consumption (14%) are used on refrigeration comprising five applications (the figures quoted below are from the Institute of Refrigeration website at http://www.ior.org.uk):

- *Domestic:* includes fridges, freezers and air conditioning (on average one fridge for a family of three).
- *Transportation:* includes refrigerated containers worldwide. There are about one million refrigerated road vehicles, half a million refrigerated containers and 1000 refrigerated ships.

- *Industrial:* includes cold store warehouses, processing plants such as dairies, brewing and ice cream, food processing and pharmaceutical.
- *Commercial:* includes supermarkets, shops, hotels and restaurants. There are about 300 000 small refrigeration systems and cold stores in the UK alone.
- *Air conditioning:* there are nearly half a million air-conditioning systems in the UK alone, in various locations such as shops, hospitals, offices, aeroplanes, trains and cars.

7.2   History of refrigeration

Records dating back to around 2000 BC indicate that people knew of the preserving effects colder temperatures had on food. Around 755 AD Khalif Mahdi provided refrigerated transport across the desert to Mecca using snow as his refrigerant, and Alexander the Great served his soldiers snow-cooled drinks around 300 BC.

Although some refrigerants have been with us for a long time, refrigeration was first recorded in 1834 when Jacob Perkins invented a vapour compression machine operated using ethyl ether.

For some decades, in the sporadic vapour compression plants that were constructed, ethyl ether remained almost the only refrigerant; it was not a good choice, the fluid is inflammable, toxic and has a normal boiling point at +39°C.

In the meantime, two other systems for producing a cold environment were invented: the air cycle machine and the absorption machine. The first came in 1844 courtesy of an American, John Gorrie, a doctor living in Florida who worked to produce ice for the comfort of patients in his hospital. It is not easy today to understand the difficulties faced by the pioneers of refrigeration some 150 years ago. Religious prejudices meant that people were against the possibility of 'producing' cold. In the middle of the 19th century, natural ice was seen as God's own ice and it was never thought that anything could replace it. Gorrie had some doubts about revealing his discovery, fearing he could be exposed to the religious censure of his fellow citizens. Accordingly he published an account in the form of a prediction of the possibility of making such a machine; this appeared in the newspaper of his city. His fear was sound, since a well-known newspaper in New York, *The Globe*, wrote shortly afterwards that 'there is a crank down in Florida that thinks that he can make ice by his machine as good as God Almighty'. Gorrie did not obtain an American patent until 1851.

The air cycle machine, despite being cumbersome and having only mediocre performance, had great success at sea. It was dominant from 1890 to 1900 but was dethroned by the CO_2 compressors. The lineage of the air machine is almost extinct today, with the exception of very high-speed compressors used for air conditioning in aircraft.

The absorption machine was invented in 1859 by F. Carré (1823–1900) who is merited with the introduction of ammonia as a refrigerant. The absorption machine was immediately successful, particularly in the USA and was clearly dominant over all other types of refrigerating machine during the period before 1875.

Compression systems truly arrived when refrigerants less dangerous than ether appeared; the first ammonia compressor was made in 1873. At the same time, sulphur dioxide, methyl chloride and carbon dioxide were introduced; the latter had great success from 1890 onwards, particularly in the field of maritime transport.

In the compression refrigerating machinery, the four aforementioned fluids, with a net prevalence of ammonia, held out until 1930. The great achievement that started in that year was the introduction as refrigerants of the halogenated hydrocarbons, first indicated by the trademark 'Freon'. Three researchers, Th. Midgley, A.L. Henne and R. McNary, working in the Frigidairie Laboratories in Dayton, Ohio, realised the suitability of these organic compounds as thermodynamic fluids in refrigeration cycles. The best candidate R12 (CCl_2F_2) was announced with appropriate fanfare at the Congress of the American Chemical Society in Atlanta, Georgia. The commercial production of R12 started in 1931 at Kinetic Chemical, Inc. in Wilmington, Delaware. In the following years until 1936, commercial production of R11, R114, R113 and R22 began.

Subsequently, many other chlorofluoro derivatives from methane and ethane were introduced and these began to be used in other technological applications outside of the refrigeration sector. Almost all fields of refrigeration have been dominated by chlorofluorocarbons; that said, ammonia remained the preferred refrigerant in large industrial machines. Freon-22 became the industry standard for commercial applications for the next four decades.

However, in 1970, chlorofluorocarbon (CFC) and hydrochlorofluorocarbon (HCFC) compounds were found to be reacting with the Earth's atmospheric ozone. In 1985, the 'ozone hole' was discovered over Antarctica. Shortly afterwards, in 1987, The Montreal Protocol put forward proposals to phase out the production and consumption of CFC and HCFC gases. In 1989, The Montreal Protocol entered into force. Since then, it has undergone seven revisions including 1990 (London), 1991 (Nairobi), 1992 (Copenhagen), 1993 (Bangkok), 1995 (Vienna), 1997 (Montreal) and 1999 (Beijing). Under The Montreal Protocol, production of CFCs was banned in 1996.

In July 2006 the F-Gas Regulation became law, covered by the Kyoto Protocol. It was derived from Regulation EC 842/2006 to reduce the risk of climate change caused by the deliberate or accidental release of refrigerant gases.

7.3 Refrigeration choice and environmental impact

The factors determining which refrigerant to use can be grouped into three sets of properties:

(1) *Thermal properties*. It can be said that the vital factor for a refrigerant to be useful is its ability to evaporate at a temperature lower than the ambient, thus absorbing heat from its surroundings, and its ability to

liquefy (apart from CO_2) at a not-too-high pressure so that it can be reused when this fluid is circulated back to the space where the cooling effect is required; just before then its pressure is reduced to the point where it can evaporate once more. The following characteristics are desirable in a refrigerant:

- The freezing point must be below the lowest temperature desired for the application.
- The latent heat of evaporation should be high (large refrigerating effect).
- The specific volume of the vapour should be small (compact machine and lower compression work needed).
- The pressure in the system should be low to reduce stresses on components and reduce the potential for leakage.
- The parameter most frequently used to indicate efficiency of refrigeration is the coefficient of performance (COP), which will be defined later, and a higher value of COP represents a superior refrigerant.

(2) *Cost consideration*. Can be cumbersome, as it is influenced by availability of fluid, manufacturing, labour costs involved and the system design as well as the efficiency of the working system, assessed by its refrigerating capacity, cost of compression energy and the coefficient of performance.

(3) *Environmental impact considerations*. These are very important these days, and they concern the refrigerant's ecological aspects such as its ozone depletion potential, global warming potential, fire/combustibility and toxicity. Environmental factors affecting the choice of refrigerant have become very important in recent years due to the increased depletion of the ozone layer and global warming caused by gases emitted into the atmosphere, including those used in refrigeration such as the CFC family. Ozone depletion potential is measured by reference to R11 (ODP = 1) and global warming potential is measured by reference to CO_2 (GWP = 1.0). It could be argued that COP is also an indicator of the environmental consideration in a refrigeration system.

For purposes of comparison, R134a has a GWP of about 1300, while that of the hydrocarbon refrigerant isobutane is about 11. Both of these refrigerants are used as replacement fluids for R12 in domestic refrigerators. The difference in GWP is significant. The contribution to global warming comes not only from the refrigerant that may leak from the system, but also from the amount of carbon dioxide that will be emitted at the power plant in producing the energy needed to run the compressor.

The effect of a refrigeration system on global warming is more accurately described by the popular concept of *total equivalent warming impact (TEWI)*, which takes into account both of these contributions to global warming. TEWI, therefore, ties in the energy efficiency of the system (and that of the power plant) in determining the effect upon global warming.

Table 7.1 Comparative properties of some common refrigerants.

Refrigerant		Condensed pressure	Theoretical COP (Relative to R22)	Ozone depletion potential	Global warming potential	Atmospheric life (years)	Toxicity
HCFC	R22	1.94	1	0.05	1700	12	Low
HFC	R134a	1.32	0.98	0	1300	14	Low
	R125	2.54	0.73	0	2800	29	Low
	R407C	2.11	0.95	0	1530	–	Low
	R410A	3.06	0.89	0	1730	–	Low
	R32	3.14	0.95	0	650	4.9	Low
Natural refrigerants	NH_3(Ammonia) R717	2.03	1.07	0	0	0	High
	C_3H_8(Propane)	1.71	0.95	0	3	10	Low
	CO_2(Carbon dioxide) R744	10.00	0.27	0	1	120	Low

7.3.1 TEWI calculation

TEWI (expressed in terms of equivalent kg CO_2 emissions) is the sum of both the direct and indirect influences and provides an index for the measure of the total warming impact of comfort conditioning and refrigeration on the environment.

TEWI = direct CO_2 emission equivalent + indirect CO_2 emission equivalent

The direct TEWI component is determined by converting refrigerant loss from a product into an equivalent CO_2 emission by multiplying the total mass of refrigerant gas emitted by its 100-year integrated GWP:

$$\text{Direct } CO_2 \text{ emission equivalent (kg } CO_2) = (\text{charge}) \times (\text{make-up rate}) \times (\text{service life}) \times (\text{GWP})$$

where:

charge = kg refrigerant in the product
make-up rate = % refrigerant charge emitted per year
service life = years of equipment operation
GWP = kg CO_2/kg refrigerant for a 100-year integration time horizon

The indirect effect is attributed to the contribution of the power requirement to drive the refrigeration unit.

$$\text{Indirect } CO_2 \text{ emission equivalent (kg } CO_2) = (\text{power}) \times (\text{operation}) \times (\text{service life}) \times (CO_2 \text{ emission from electricity generation})$$

where:

power = equipment power requirements, kW
operation = hours of equipment operation per year (equivalent full load)
service life = years of equipment operation
CO_2 emission from electricity generation = kg CO_2 emitted per kWh generated

Table 7.1 shows a comparison of the properties of some of the refrigerants in common use.

7.4 Refrigeration system components

A simple refrigeration system (shown in Figure 7.1) consists of four basic units: the compressor, the expansion valve, the condenser and the evaporator.

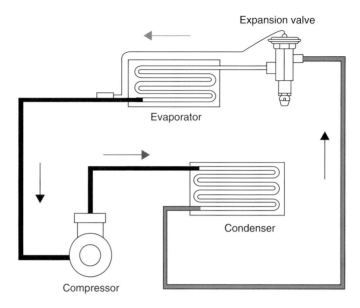

Figure 7.1 Basic refrigeration system. Dark grey line indicates gas; pale grey line indicates liquid.

7.4.1 The compressor unit

There are three types of compressor; each having advantages and disadvantages in respect of cost and performance:

- Reciprocating compressor. To date, this is the most common type used in refrigeration.
- Centrifugal compressor.
- Rotary compressor (screw and scroll).

Although both centrifugal and rotary compressors have been used in many applications, they are limited to small compression ratio applications; they have the advantage over piston compressors in terms of having higher efficiency at low-pressure ratios and lower noise. In this text, we will focus on reciprocating compressors only.

A reciprocating compressor consists of a piston and cylinder; the refrigerant vapour enters and leaves at the top of the cylinder through inlet and outlet valves. These valves are usually spring loaded and their alternate open/close positions are controlled by differential pressure.

The function of the compressor in a refrigeration system is to draw in the relatively low-pressure refrigerant gas from the evaporator unit and compress it to a higher pressure. This enables the refrigerant vapour to be condensed back into liquid (the actual pressures are dictated by the saturation temperatures for the refrigerant used).

The compressor unit is usually driven by an electric motor, mostly hermetic and semi-hermetic and built into a common casing.

When selecting a compressor, it is important to check the manufacturer's data to select the minimum power input machine which can produce the required cooling effect.

In order to estimate the power input to the compressor, it is important to explain that there are three sets of effects which are encountered in delivering the compression work to the refrigerant:

- Electrical efficiency of the motor; a motor power of 20 kW may, in fact, involve 22 kW of actual mains input of electrical energy. Typically, electrical efficiency is much better than 90%.
- Mechanical efficiency in driving the compressor; this is due to linkage losses between the motor and the piston. System design will dictate the value of mechanical efficiency.
- Volumetric efficiency due to the fact that the actual volume flow of refrigerant is less than the ideal swept volume flow through the cylinder. There are three main reasons for this: pressure drop across the valves (known as wiredrawing), high pressure 'blow-by' past the piston-cylinder and re-expanding dead volume.

Volumetric efficiency for a given refrigerant is greatly affected by the compression action. It decreases linearly with the compression ratio. For R12 it decreases at the rate of 6% for a unit increase in pressure ratio. Similar behaviour is expected for other refrigerants (manufacturers of refrigerants will usually supply such data upon request).

The volume flow rate of the refrigerator in the compressor indicates the physical size of that compressor. The greater its magnitude, the larger the displacement of the compressor must be. The piston displacement of a reciprocating compressor is the total cyclinder volume swept through by the piston in any given time interval. For a single-acting compressor:

$$V_f = \dot{V}_t = \frac{\pi}{4} D^2 L N n \qquad\qquad [7.1]$$

where

V_f is the theoretical volume flow rate
D is the bore or diameter of the compressor cylinder
L is the length of stroke
N is the speed of the compressor
n is the number of cyclinders (multiply by 2 if double acting)

The actual volume of vapour displaced by the compressor is somewhat smaller than that quoted above. It is given by:

$$\dot{V}_a = \dot{V}_t \cdot \eta_v \qquad\qquad [7.2]$$

where η_v is the volumetric efficiency which is a function of the compression ratio.

The mass flow rate circulated in the system is equal to the product of the actual volume flow rate and the density (the reciprocal of specific volume) of the suction vapour at the compressor inlet:

$$\dot{m} = \dot{V}_a / v_1 \qquad\qquad [7.3]$$

where v_1 is the specific volume at entry to the compressor.

7.4.2 The expansion valve

This is the smallest process component of the refrigerant unit, yet it has an important function in the cycle. In principle, the expansion valve allows the liquid refrigerant to pass from a small section to a large space. This causes the warm liquid to partially vaporise. Since there is a predetermined drop in the pressure of the refrigerant, the corresponding saturation temperature of the refrigerant is lower after expansion. A well-insulated expansion process is one in which the enthalpy of the refrigerant remains constant.

There are three types of expansion valve, categorised in relation to the mechanism of adjustment: pressure expansion valves (PEVs); thermostatic expansion valves (TEVs); and electronic expansion valves operated by a stepper motor (EEVs).

Pressure expansion valves (PEVs)

The constant pressure valve maintains a constant pressure on the low side. The pressure-balancing mechanism is shown in Figure 7.2.

In operation, the valve feeds enough liquid refrigerant to the evaporator to maintain a constant pressure in the coils. This type of valve is generally used

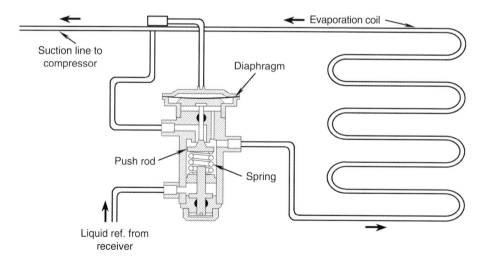

Figure 7.2 Pressure expansion valve (PEV).

in a system where constant loads are expected. When a large variable load occurs, the valve will not feed enough refrigerant to the evaporator under high load and will overfeed the evaporator at low load.

Thermostatic expansion valves (TEVs)

In this type, the refrigerant pressure is dropped through an orifice, and the flow of refrigerant is regulated by a needle valve and a diaphragm arrangement, as shown in Figure 7.3. The diaphragm is actuated by the pressure inside the controlling phial, which senses the temperature of the refrigerant leaving the evaporator, keeping it at about 5°C above the evaporation temperature to ensure that the refrigerant is in a superheated gaseous state for the safe operation of the compressor. This temperature difference can be adjusted at will, and is usually set between 4 and 7°C superheat; once set, the mechanism will automatically operate the flow of refrigerant at the pre-set superheat temperature difference. If the load changes, the temperature of the refrigerant leaving the evaporator will also change; the mechanism will automatically adjust the refrigerant flow to accommodate the load change.

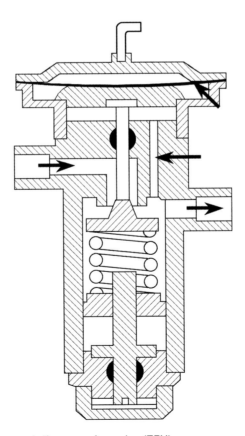

Figure 7.3 Thermostatic expansion valve (TEV).

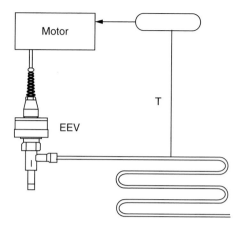

Figure 7.4   Electronic expansion valve (EEV).

Electronic expansion valves (EEVs)

A motorised refrigerant valve has a spool in an axially moveable valve body for controlling flow from a high-pressure inlet for expansion and discharge at substantially reduced pressure to an outlet in the valve body. A ball screw is provided on the remote end of the spool and the ball screw is driven by a rotating shaft of a stepper motor attached to the valve body to provide fine resolution for linear movement of the spool in the valve passage. A separate isolated passage through the valve body is connected to receive evaporator discharge flow; this has a thermistor sensing temperature and provides a temperature signal to an electronic controller for providing a driver signal to the stepper motor. The stepper motor and ball screw drive to the spool provide the desired modulation of refrigerant flow in response to the signal from the controller (Figure 7.4).

7.4.3   The condenser

This unit in the refrigeration system has the function of liquefying the refrigerant vapour. It consists of a nest of tubes through which the refrigerant flows (Figure 7.5); on the outside, a coolant medium is allowed to remove the heat from the refrigerant, hence allowing it to condense into liquid.

There are three types of condenser in use in air-conditioning and refrigeration applications:

- air-cooled;
- water-cooled;
- evaporative type.

A typical household refrigerator is air-cooled; free convection can be enhanced by the use of fins on the outer surface of condenser tubes.

A well-designed air-cooled condenser operates with a condensing temperature no higher than 14°C above the ambient temperature. Air-cooled condensers may be assisted by an air-fan.

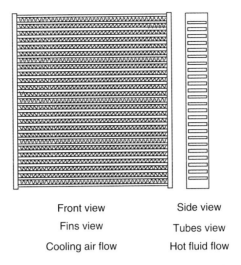

Front view   Side view

Fins view   Tubes view

Cooling air flow   Hot fluid flow

Figure 7.5   Condensers and evaporators (heat exchangers).

Water-cooled condensers are of the shell-and-tube type or the brazed plate cross-flow heat exchanger type. The cooling water flows in tubes inside the shell, and the refrigerant occupies the shell. A good system operates with a temperature rise of about 5K for water flowing through the condenser, and 5K temperature difference between the condensing refrigerant and that of the water leaving the condenser.

In evaporative condensers, the refrigerant is condensed in tubes which are wetted by water and over which air is forced. This type of condenser operates with similar temperatures to water-cooled types.

7.4.4   The evaporator

The evaporator is represented as a surface enclosure which cools the inside air of a refrigerated space. There are two types of evaporator: direct expansion evaporators and flooded evaporators, and the design of an evaporator is dictated by its specific application.

Direct expansion evaporators

These consist of a matrix containing a number of tubes with the saturated liquid entering from one end and leaving as a vapour at the opposite end due to heat absorbed from a warmer load surrounding the refrigerant tubes. Finned surfaces are used to maximise heat transfer rates. Small commercial and domestic refrigerators are of the direct type.

Flooded evaporators

Evaporators are divided into a number of blocks or circuits in order to limit the frictional pressure drop of the refrigerant flowing. It is important that each circuit has the same length and that they are all connected to a common header (Figure 7.5).

Flooded evaporators work on the principle that when liquid is poured into a comparatively large passage or vessel, it evaporates, and the refrigerant vapour is drawn off the top of the shell. Heat exchange is enhanced when the liquid refrigerant floods the surface of the coil.

7.5 Heat pump and refrigeration cycles

7.5.1 The heat engine

In heat engines (see Figure 7.6), heat is received by the working fluid at a high temperature (hot reservoir) and is then rejected with the exhaust gases at a lower temperature (cold reservoir). While the working fluid expands under heat, it will move the piston, hence rotating the shaft, and the opposite happens at the cold reservoir - the cooling contracts the volume of the working gas, hence there is a continued rotation of the crank shaft and a net amount of work is produced, as demonstrated by the continued circular motion of the fly wheel, which can be imagined as a car wheel.

The efficiency of such a cycle is defined as the ratio of output mechanical work (energy) and the energy input as heat converted from the combustion of the fuel:

$$\eta_{HE} = \frac{W}{Q_1}$$ [7.4]

The energy balance of the system in Figure 7.6 implies that the net work output is equal to the difference between the energy supplied from the fuel and the energy lost with the exhaust, hence:

$$W = Q_1 - Q_2$$ [7.5]

$$\therefore \eta_{HE} = \frac{Q_1 - Q_2}{Q_1}$$ [7.6]

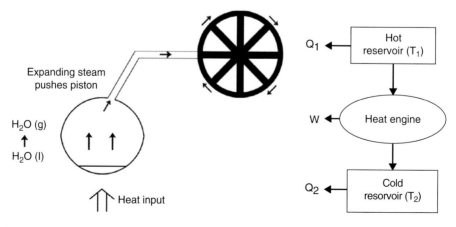

Figure 7.6 Heat engine.

But the quantity of heat associated with the body temperature of the system's gases is given by:

$$Q = mC_p T \qquad\qquad\qquad\qquad\qquad [7.7]$$

$$\therefore \eta_{HE} = \frac{T_1 - T_2}{T_1} \qquad\qquad\qquad\qquad\qquad [7.8]$$

This is known as *Carnot efficiency*, which is the maximum possible value that can be achieved in any situation.

7.5.2 Reversed heat engine (heat pump/refrigerator)

It is possible to reverse the operation of the above system (heat engine) only if external work is supplied to the machine; it will then operate by rejecting heat to a hot reservoir. This is known as a heat pump. Alternatively, the cold reservoir provides a source of a relatively cold environment well below the ambient air temperature and hence is frequently used as a refrigerator.

The term 'heat pump' is given to a machine whose principal function is to supply heat at an elevated temperature, whereas the term 'refrigerator' is given to a machine whose function is the extraction of heat from a cold space. It is common to see a single machine with a dual function, working as both a heat pump and a refrigerator (Figure 7.7). For example, one machine can provide a cold space for food storage and also supply heat to a domestic hot water tank simultaneously.

The criterion of performance of the cycle, expressed as the ratio of output/input, depends upon what is regarded as output. If this is considered to

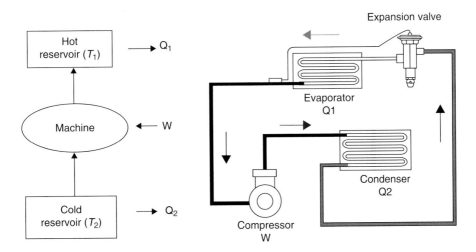

Figure 7.7 Heat pump/refrigerator. Dark grey line represents gas; pale grey line represents liquid.

be heat, then the efficiency definition used with heat engines is not valid for refrigerator/heat pumps.

Another term is, therefore, used to express the effectiveness of a refrigerator/heat pump and is known as the *coefficient of performance (COP)*. This is defined as the ratio of heat output to the work input.

Hence, there are two different expressions for COP - one for the refrigerator and a second for a heat pump:

$$COP_R = \frac{Q_2}{W} \qquad\qquad [7.9]$$

$$COP_{HP} = \frac{Q_1}{W} \qquad\qquad [7.10]$$

But the first law of thermodynamics can be written in terms of Q_1 as:

$$Q_1 = Q_2 + W \qquad\qquad [7.11]$$

dividing by W gives:

$$\frac{Q_1}{W} = \frac{Q_2}{W} + 1 \qquad\qquad [7.12]$$

Or

$$(COP)_{HP} = (COP)_R + 1 \qquad\qquad [7.13]$$

Since $Q = mC_pT$, it is customary to see the coefficient of performance written in terms of temperatures:

$$\therefore (COP)_{HP} = \frac{T_1}{T_1 - T_2} \qquad\qquad [7.14]$$

For a refrigerator:

$$(COP)_R = \frac{T_2}{T_1 - T_2} \qquad\qquad [7.15]$$

Note that

$$COP_{HP} = \frac{1}{\eta_{HE}} \qquad\qquad [7.16]$$

A comparison of a heat engine, heat pump and a refrigerator is shown in Figure 7.8.

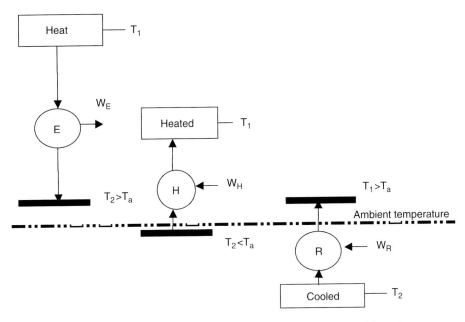

Figure 7.8 Comparison between a heat engine, a heat pump and a refrigerator.

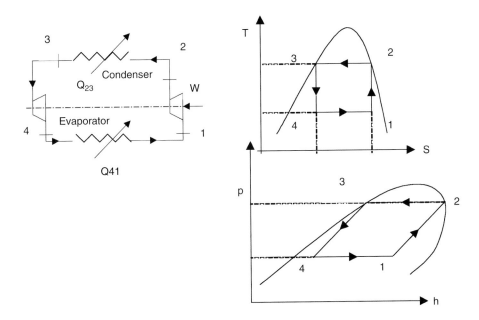

Figure 7.9 Ideal refrigeration cycle.

7.5.3 Carnot refrigeration cycle

The layout of a refrigerator operating on a reversed Carnot cycle is shown in Figure 7.9. The four processes in the reversed Carnot cycle are as follows:

• Vapour is compressed isentropically from 1 to 2.
• The vapour is condensed at constant pressure (and temperature) from 2 to 3.

- The fluid is then expanded isentropically from 3 to 4.
- The fluid is evaporated at constant pressure from 4 to 1.

Based on a unit mass flow of refrigerant in the system of Figure 7.9:

Refrigeration effect $= h_1 - h_4$ [7.17]

Compressor work input $= h_2 - h_1$ [7.18]

$$\therefore COP = \frac{h_1 - h_4}{h_2 - h_1}$$ [7.19]

7.5.4 Simple refrigeration cycle

The work obtained in the expander can be used to reduce the external work required to drive the compressor. However, the greater initial cost and mechanical complication of the expander are not justified in practice and a major simplification is obtained by replacing the expander with a simple throttle valve. The evaporator is also modified (its size is slightly increased to ensure that the refrigerant heading to the compressor is definitely in a dry vapour state). This leads to a cycle which has a lower COP but which has other advantageous features.

The new cycle, shown in Figure 7.10, consists of the following processes:

- Isentropic compression from 1 to 2.
- Condensation at constant pressure from 2 to 3.
- Throttling at constant enthalpy ($h_3 = h_4$) from 3 to 4.
- Evaporation at constant pressure from 4 to 1.

For this cycle, the performance is calculated for a unit mass of refrigerant, as follows:

Refrigerating effect $Q_{41} = h_1 - h_4$

Work done $W_{12} = h_2 - h_1$

$$COP_R = \frac{Q_{41}}{W_{12}} = \frac{h_1 - h_4}{h_2 - h_1}$$

7.5.5 Practical refrigeration cycle

The simple cycle shown in Figure 7.10 can be modified further to improve performance by any or all of the following: superheating; subcooling; use of a heat exchanger.

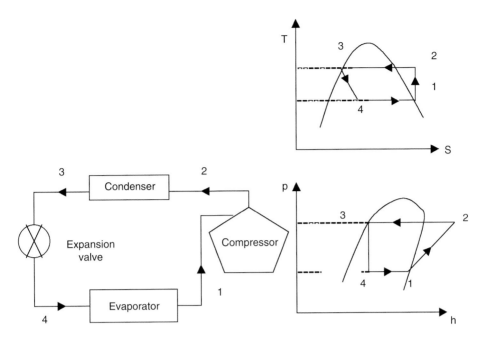

Figure 7.10   Simple refrigeration cycle.

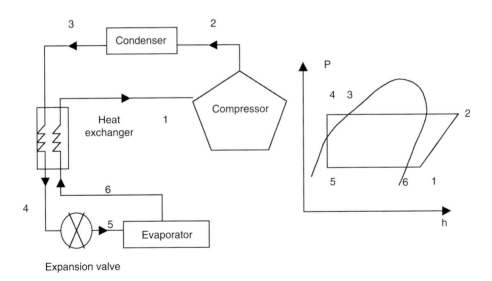

Figure 7.11   Refrigerator with a heat exchanger.

Superheating

In the simple Carnot cycle, the refrigerant enters the compressors as vapour (condition 1 in Figure 7.10). The refrigerant leaving the evaporator is super-heated to eliminate the possibility of wet refrigerant entering the compressor (Figure 7.11, process 6 to 1).

Subcooling

A further modification is the subcooling of the liquid refrigerant in the condenser (i.e. the temperature is reduced below the saturated liquid temperature at the upper pressure). The final temperature is usually below the temperature of the coolant in the condenser (Figure 7.11, process 3 to 4).

Both superheating and subcooling increase the refrigeration effect, resulting in an improved coefficient of performance.

The use of a heat exchanger

The use of a liquid-to-suction heat exchanger subcools the liquid from the condenser with suction vapour coming from the evaporator. The arrangement is shown in Figure 7.11. The effect of using a heat exchanger results in producing a double effect of superheating and subcooling simultaneously.

7.5.6 Irreversibilities in the refrigeration cycle

Pressure drop in the refrigeration cycle

The actual vapour-compression cycle suffers from pressure drops in the condenser and evaporator units due to friction. As a result of this pressure drop, the compression process requires more work than in the standard cycle (see Figure 7.12).

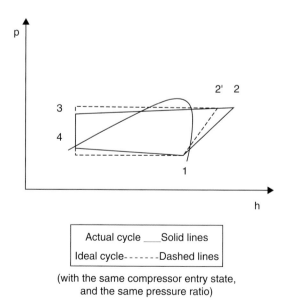

| Actual cycle ___Solid lines |
| Ideal cycle- - - - - - -Dashed lines |

(with the same compressor entry state, and the same pressure ratio)

Figure 7.12 Irreversibilities in the refrigeration cycle.

Irreversible compression

Due to irreversibilities in the compressor, there is an increase in the entropy of the working fluid as it goes through the compressor. A compressor works with an isentropic efficiency given by (see Figure 7.12):

$$\eta_{ic} = \frac{h'_2 - h_1}{h_2 - h_1}$$ [7.20]

where:

h_1 is the enthalpy at the compressor inlet condition

h_2' is the enthalpy at the ideal compressor (100% isentropic) outlet condition

h_2 is the enthalpy at the actual compressor (not 100% isentropic) outlet condition.

7.5.7 Multi-stage compression

The advantage of compressing in more than one stage is that it reduces input work. There are two ways in which this can be achieved: compression using only an intercooler and compression using flash intercooling.

Two-stage compression with intercooler only

The vapour leaves the low-pressure compressor (Figure 7.13, point 2), is cooled to the saturated condition (point 3), then further compressed in the high-pressure cylinder (point 4), and from there to point 1 the sequence is the same as in the simple plant.

Refrigeration effect $= (h_1 - h_6)$ [7.21]

Compressor's work $= (h_2 - h_1) + (h_4 - h_3)$ [7.22]

$$COP_R = \frac{(h_1 - h_6)}{(h_2 - h_1) + (h_4 - h_3)}$$ [7.23]

Two-stage compression with flash intercooling

For 1kg/s of fluid passing through the HP compressor, let $\dot{m}$ kg/s pass through the evaporator.

The vapour leaves the high-pressure cyclinder at point 4 (see Figure 7.14) generally in the superheated state, and passes to the condenser. From the

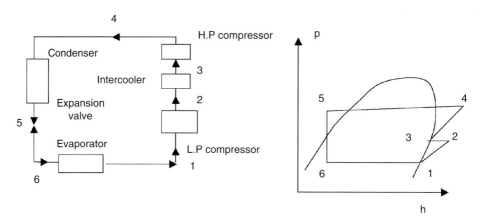

Figure 7.13 Two-stage compression with intercooler.

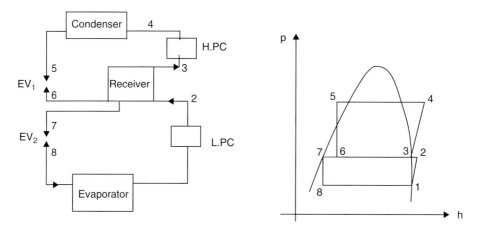

Figure 7.14 Two-stage compression with flash intercooling.

condenser at point 5, it passes to the first expansion valve V_1 (process 5 to 6) and is discharged into the receiver or flash chamber. At point 7, $\dot{m}$ kg/s of liquid leave the chamber and pass through the second expansion valve V_2 (process 7 to 8), subsequently flowing into the evaporator. The vapour, which may be wet, dry saturated or superheated leaving the evaporator, is shown dry saturated in Figure 7.14. It is taken in by the low-pressure cylinder and, after compression, it is discharged, in a superheated condition, into the receiver at point 2. Here it mixes with the liquid-vapour mixture which entered the receiver after the first throttling, and is cooled to the dry saturated condition.

Thus, cooling is effected by a heat transfer to the liquid, and this causes a portion of the liquid to flash evaporate. At point 3, 1kg/s of dry saturated vapour is drawn into the high-pressure cyclinder and compressed to the condenser pressure. The cycle is thereby completed.

Consider the energy conservation in the receiver:

$$1 \times h_6 + \dot{m} \times h_2 = 1 \times h_3 + \dot{m} \times h_7$$

Hence

$$\dot{m} = \frac{h_3 - h_6}{h_2 - h_7} \qquad [7.24]$$

Consider the flow of 1 kg/s through the condenser:

$$RE = \dot{m}\,(h_1 - h_8) \qquad [7.25]$$

$$W_c = \dot{m}\,(h_2 - h_1) + (h_4 - h_3) \qquad [7.26]$$

$$COP = \frac{\dot{m}\,(h_1 - h_8)}{\dot{m}\,(h_2 - h_1) + (h_4 - h_3)} \qquad [7.27]$$

7.5.8   Multipurpose refrigeration systems with a single compressor

Some applications require refrigeration at more than one temperature. This could be accomplished by using a separate throttling valve and a separate compressor for each evaporator operating at different temperatures. However, such a system would be bulky and probably uneconomical. A more practical and economical approach would be to route all the exit streams from the evaporators to a single compressor and to let it handle the compression process for the entire system.

Consider, for example, an ordinary refrigerator-freezer unit. A simplified schematic of the unit and the T-s diagram of the cycle are shown in Figure 7.15. Most refrigerated goods have a high water content, and the refrigerated space must be maintained above the ice-point (at about 5°C) to prevent freezing.

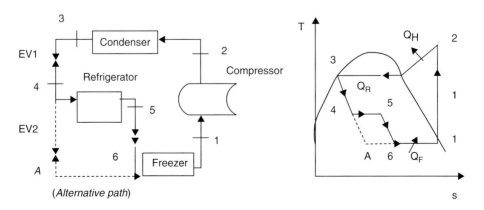

Figure 7.15   Fridge-freezer.

The freezer compartment, however, is maintained at about –15°C. Therefore, the refrigerant should enter the freezer at about –25°C to have heat transfer at a reasonable rate in the freezer. If a single expansion valve and evaporator were used, the refrigerant would have to circulate in both compartments at about –25°C, which would cause ice formation in the neighbourhood of the evaporator coils and dehydration of the produce. This would not be acceptable to a household. This problem can be eliminated by throttling the refrigerant to a higher pressure (and hence temperature) for use in the refrigerated space and then throttling it to the minimum pressure for use in the freezer. All of the refrigerant leaving the freezer compartment is subsequently compressed by a single compressor to the condenser pressure.

7.6 Worked examples

Worked example 7.1

A heat pump used for heating/cooling a building has a COP_r of 4. During the cooling season, the rate of heat removal is 16 kW. Determine:

(a) The energy input to the compressor.
(b) The winter heating capacity of this system.

Solution:

(a) Since the coefficient of performance is known,

$$COP_r = 4 = \frac{Q_e}{W_c} = \frac{16}{W_c}$$

Hence, the work input to the compressor can be determined from the above definition:

$$W_c = 16/4 = 4\,\text{kW}$$

(b) In order to determine the heating capacity, remember the COP relationship for heating and cooling modes of a heat pump:

$$COP_h = COP_c + 1 = 4 + 1 = 5$$
$$COP_h = \frac{Q_h}{W_c} = \frac{Q_h}{5}$$
$$\therefore Q_h = 5 \times 4 = 20\,\text{kW}$$

Worked example 7.2

A refrigerator uses R12 as the working fluid. The refrigerant enters the compressor as saturated vapour at 2 bar and on leaving the condenser it is saturated liquid at 8 bar (absolute). Assume that the compression process is 100% isentropic, that there are no superheating or subcooling effects and that pressure losses are negligible. Calculate the work of compression, the refrigerating effect and the coefficient of performance. Compare this value with the Carnot performance.

Solution:

The system is shown in Worked example 7.2, Figure 1.

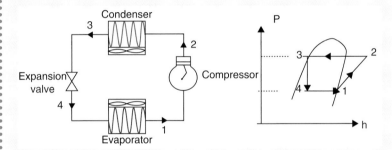

Worked example 7.2, Figure 1

Construct the cycle on the *P-h* chart. Find:

$$h_1 = 245 \, \text{kJ/kg}$$
$$h_2 = 270 \, \text{kJ/kg}$$
$$h_3 = h_4 = 130 \, \text{kJ/kg}$$

Hence:

Work of compressor $\quad W_c = h_2 - h_1 = 25 \, \text{kJ/kg}$
Refrigeration effect $\quad RE = h_1 - h_4 = 115 \, \text{kJ/kg}$

Coefficient of Performance $\quad COP_R = RE/W_c = 4.6$

$$\text{Ideal COP} = \frac{T_2}{T_1 - T_2} = \frac{260}{305 - 260} = 5.77$$

The ideal COP value is higher than COP_R, as expected.

Worked example 7.3

A refrigerator is using R134a as a retrofit working fluid to replace R12 in the system of Worked example 7.2. The refrigerant enters the compressor as saturated vapour at 2 bar and, on leaving the condenser, it is saturated liquid at 8 bar (absolute). Assume that the compression process is 100% isentropic, that there are no superheating or subcooling effects and that pressure losses are negligible. Calculate the work of compression, the refrigerating effect and the coefficient of performance.

Solution:

Construct the cycle on the P-h chart, find enthalpies at key state points:

$h_1 = 392\,kJ/kg$
$h_2 = 420\,kJ/kg$
$h_3 = h_4 = 244\,kJ/kg$

Hence:

Work of compressor	$W_c = h_2 - h_1 = 28\,kJ/kg$
Refrigeration effect	$RE = h_1 - h_4 = 148\,kJ/kg$
Coefficient of Performance	$COP_R = RE/W_c = 5.28$

which is better than that for R12 (4.6).

The simulation of this example was reproduced on a software package and the results are shown in Worked example 7.3, Figure 1. Refrigerant: R134a data:

$T_e\,[°C] = -10.09$ (2 bar)
$T_c\,[°C] = 31.33$ (8 bar)
DT subcooling [K] = 0.00
DT superheat [K] = 0.00
Isentropic efficiency = 1.00

Calculated:
$Q_e\,[kJ/kg] = 147.871$
$Q_c\,[kJ/kg] = 176.478$
$W\,[kJ/kg] = 28.607$
$COP\,[-] = 5.17$

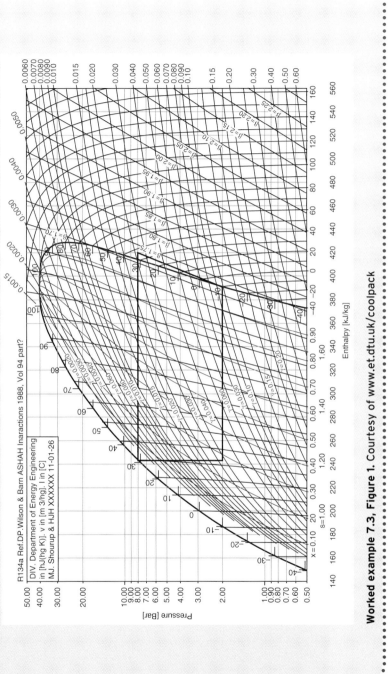

Worked example 7.3, Figure 1. Courtesy of www.et.dtu.uk/coolpack

Worked example 7.4

The refrigeration plant in Worked example 7.2 has a double-acting, reciprocating compressor (a bore of 250 mm, a stroke of 300 mm) running at a speed of 200 rev/min, with a volumetric efficiency of 85%. Calculate:

(a) The mass flow rate of refrigerant.
(b) The required power input to the electric motor when the mechanical efficiency is 94% and the electrical efficiency is 96%.

Solution:

(a) At the compressor inlet, the specific volume is found from the chart to be:

$$v_1 = 0.08 \, \text{m}^3/\text{kg}$$

The swept volume of the compressor is:

$$V_s = \frac{\pi}{4}D^2L = \frac{\pi}{4}(0.25)^2 \times 0.3 = 0.0147 \, \text{m}^3/\text{cycle}$$

The mass flow of refrigerant is therefore:

$$\dot{m}_f = \frac{2 \times n \times N \times \eta_v \times V_s}{v_1}$$

$$\dot{m}_f = \frac{2 \times 1 \times 200 \times 0.85 \times 0.0147}{0.08 \times 60} = 1.043 \, \text{kg/s}$$

(b) Taking the enthalpy values from Worked example 7.2:

$$\text{Power} = \frac{\dot{m}(h_2 - h_1)}{\eta_m \eta_e} = \frac{1.043(270 - 246)}{0.94 \times 0.96} = 27.7 \, \text{kW}$$

Worked example 7.5

A refrigeration plant running on R134a operates between saturation temperatures of 0 and 45°C and has a refrigerating capacity of 1MW. The vapour is dry saturated on entry to the compressor and there is no subcooling in the condenser. Assume that the isentropic efficiency for the compressor is 80%, the motor efficiency is 90% and the mechanical efficiency is 90%. Calculate the coefficient of performance of the plant and the required power input to the compressor.

Solution:

Refer to Worked example 7.5, Figure 1.

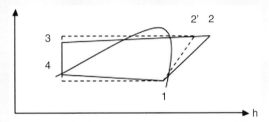

Worked example 7.5, Figure 1

Enthalpy values are extracted from a refrigerant table: $h_1 = 299\,$kJ/kg, $h_3 = 165$kJ/kg.

For 100% isentropic compression, the ideal compressor outlet enthalpy is estimated as: $h_{2s} = 326$kJ/kg

The actual outlet enthalpy from the compressor is determined using the isentropic efficiency given by:

$$\eta_{ic} = \frac{h_{2'} - h_1}{h_2 - h_1} \xrightarrow{\text{hence}} h_2 = h_1 + \left(\frac{h_2 - h_1}{\eta_{ic}} \right) \quad \text{or}$$

$$h_2 = 299 + \frac{326 - 299}{0.8} = 332.75\text{kJ/kg}$$

Therefore:

$$COP_{ref} = \frac{Q_{41}}{W_{12}} = \frac{h_1 - h_4}{h_2 - h_1} = \frac{299 - 165}{332.75 - 299} = 3.97$$

Refrigerating effect $= Q_{41} = \dot{m}_R(h_1 - h_4) = \dot{m}_R\,(299 - 165)$
$= 1000\,$kW (given)

i.e. mass flow rate $= \dot{m}_R = \dfrac{Q_{41}}{h_1 - h_4} = 7.46 \, \text{kg/s}$

Then:

$$\text{Power} = \dfrac{\dot{m}(h_2 - h_1)}{\eta_m \eta_e} = \dfrac{7.46(332.75 - 299)}{0.9 \times 0.9} = 311 \, \text{kW}$$

Worked example 7.6

It is proposed to use a heat pump working on the ideal vapour-compression cycle for the purpose of heating the air supply to a building. The supply of heat is taken from a river at 7°C. Air is to be delivered into the building at 1.013 bar and 20°C at a rate of 0.5 m³/s. The air is heated at constant pressure from 10 °C as it passes over the condenser coils of the heat pump. The refrigerant is R134a, which is dry saturated leaving the evaporator; there is no subcooling in the condenser. A temperature difference of 17K is necessary for the transfer of heat from the river to the refrigerant in the evaporator. The delivery pressure of the compressor is 11.545 bar. Calculate the mass flow of refrigerant and the motor power required to drive the compressor if the overall electromechanical efficiency is 85%; also determine the coefficient of performance for heating, COP_{HP}.

Solution:

The saturation temperature in the evaporator is given by $7 - 17 = -10°C$. Therefore:

$h_1 = 288.86 \, \text{kJ/kg}$ and $s_1 = 1.7189 \, \text{kJ/kg K}$

Since $s_2 = s_1$,

$$h_2 = 319.54 + \dfrac{(1.7189 - 1.7028)}{(1.7440 - 1.7028)}(332.87 - 319.54) = 324.75 \, \text{kJ/kg}$$

$h_3 = 162.93 \, \text{kJ/kg}$

Mass flow rate of air

$$\dot{m}_a = \dfrac{PV}{RT} = \dfrac{0.5 \times 1.013 \times 10^5}{287 \times 293} = 0.60 \, \text{kg/s}$$

And

> Heat required = mass × specific heat capacity × temperature difference.

Since

$$Q = m_a\, Cp\, \Delta T = 0.60 \times 1.005\,(20 - 10) = 6\,kW$$

Also

$$Q = m_R\, \Delta h$$

i.e. mass flow rate of refrigerant

$$\dot m_R = \frac{Q}{(h_2 - h_3)} = \frac{6.0}{324.75 - 162.93} = 0.037\,kg/s$$

$$\text{Motor power } W_{12} = \frac{m_R(h_2 - h_1)}{\eta_m \eta_e} = \frac{0.037(324.75 - 288.86)}{0.85} = 1.56\,kW$$

$$COP_{HP} = \frac{Q_e}{W_c} = \frac{6.0}{1.56} = 3.8$$

Worked example 7.7

A heat pump rated at 100 kW heating capacity operates on R134a with an evaporator temperature of –10°C and a condensing temperature of 50°C. The system has a 10K suction superheat and 2K liquid subcool. The compressor has a displacement of 0.1 m³/s and a volumetric efficiency of 70%; the motor-drive has an overall efficiency of 80%. If the electricity costs 5p/kWh, determine the cost of heating using this plant.

Solution:

The state points are marked on the R134 P-h chart; the following values are read off the chart:

$v_1 = 0.1\,m^3/kg$
$h_1 = 305\,kJ/kg$
$h_2 = 340\,kJ/kg$
$h_3 = 170\,kJ/kg$

$$\dot{m} = \dot{V}_s \eta_v / v_1 = \frac{0.1 \times 0.70}{0.1} = 0.70 \, \text{kg/s}$$

Work of compressor $W_{12} = \dfrac{m_R(h_2 - h_1)}{\eta_m \eta_e} = \dfrac{0.7(340 - 305)}{0.80} = 30.6 \, \text{kW}$

Heat transfer $Q_c = \dot{m}(h_2 - h_3) = 0.70(305 - 170) = 94.5 \, \text{kW}$
Cost of compression $= 30.6 \times £0.06 = £1.84$

Cost/kW heating $= \dfrac{£1.84}{94.50} = 1.94\text{p/kW}$

Which is one-third the cost of heating by a 100% electrical heater! This example demonstrates the efficiency of using a heat pump for heating purposes.

7.7 Tutorial problems

7.1 A heat pump operates between 2 and 8 bar (absolute pressures) and uses R12 as the working fluid. The refrigerant enters the compressor as dry saturated vapour and is compressed isentropically (100%) then condensed at constant pressure, leaving the condenser as saturated liquid, expanded to the lower pressure of the evaporator where it evaporates at constant pressure before returning to the compressor to complete the cycle. Determine the coefficient of performance.
Ans. (5.6)

7.2 Determine the effect of an 80% efficiency of the compressor on the value of the coefficient of performance for the system in Question 7.1.
Ans. (4.7)

7.3 A heat exchanger is fitted to the heat pump in Question 7.1 above, resulting in the refrigerant being superheated by 10 degrees before entering the compressor. This is accompanied by a subcooling of 10 degrees at the condenser outlet. Determine the new coefficient of performance.
Ans. (6.4)

7.4 The refrigerator in Question 7.1 is retrofitted with R134A in order to comply with the environmental impact regulations. Assume that the system will operate under the same pressure conditions, and use the *P-h* chart of R134A to determine:

- the work of the compression;
- the refrigerating effect;
- the heating effect;
- the coefficient of performance for cooling and for heating.

Ans. (35 kJ/kg, 150 kJ/kg, 185 kJ/kg, 4.29, 5.29)

7.5 A heat pump with a COP_R of 4 is used to heat/cool a building. During the cooling season, the rate of heat removal is 16 kW. Determine:

- the winter heating capacity of this system;
- the energy input to the compressor.

Ans. (20 kW, 4 kW)

7.8 Case Study: Star Refrigeration Ltd - heat pumps in a chocolate factory. May 2010, UK

Due to their effect on the climate, HCFC refrigerants were recently phased out and replaced by HFCs, such as R134a, R407C and R404A. Although these refrigerants have no Ozone Depletion Potential (ODP), they have a high Global Warming Potential (GWP), in excess of 1000. As refrigerant leakage happens throughout the life cycle of the equipment, the use of these refrigerants is having a negative impact on global warming. Natural refrigerants are thus preferred as the post-HFC refrigerant. Among them, carbon dioxide (CO_2) and ammonia are natural substances which have been used in the past and are now making a comeback. However they do have some problems such as toxicity, and this is resolved by housing the plant outside occupied areas. Ammonia is neutral as far as ODP and GWP making it the ideal solution. See Case Study 7.8, Table 1.

The case study presented here corresponds to a real application completed in 2010 for a chocolate factory in England.

Case Study 7.8, Table 1   Properties of common and natural refrigerants.

Type	Refrigerant	ODP	GWP	Flammability	Toxicity
HCFC	R22	0.055	1700	No	No
HFCs	R134a	0	1300	No	No
	R407C	0	1530	No	No
	R404A	0	3260	No	No
Natural	CO2	0	1	No	No
refrigerants	NH3	0	~0	No	Yes

Star Collaboration Gives Free Hot Water

Customer:	Chocolate factory in England
Location:	England
Equipment:	Neatpump
Refrigerant:	Ammonia
Capacity	1.25 MW Heating
	3.20 MW Cooling
Temperature	–5°C to 60°C

Star Refrigeration has been providing the factory with refrigeration solutions since 1990 and this strong partnership led to the latest challenge for Star. The challenge was to find a heat pump solution for reducing the factory's refrigeration and heating energy demands.

The team had previously completed an energy audit on their central coal-fired boilers, the steam distribution and all of the end user heating systems throughout the factory. This enabled the team to clearly identify, grade and consolidate the various end user heating requirements which identified significant design and operational inefficiencies.

Previously one central coal-fired steam generation plant served all of the individual end users, where high grade steam would be degraded to suit the processes. The new concept was to simply heat the water to the desired process temperature and the Star Neatpump (Case Study 7.8, Figure 1) would serve to provide hot water to end users requiring 60°C and to preheat those operating in excess of 60°C.

The company's global commitment to reducing the environmental impact of its operation demands the use of natural refrigerants for all new factory process refrigeration equipment. This presented a number of challenges for Star as heat pumps until now had been either HFC dependent, which was not an option, or they had utilised reciprocating compressors which were beginning to show high regular maintenance costs or with screw compressors operating at their limit.

Case Study 7.8, Figure 1   Star Refrigeration's Neatpump.

Star Refrigeration, Vilter Manufacturing Inc (USA) and Cool Partners (a Danish consultancy) formed a collaborative effort to devise a high-pressure heat pump solution using ammonia and screw compressors up to 90°C. This enabled Star to comfortably take heat from the 0°C process glycol at −5°C and lift it to 60°C in one stage for heating.

Their challenge was to utilise a heat pump that drew electrical energy and thereby reduce the load on the proposed replacement gas-fired equipment. Alternative solutions such as Geothermal and CHP were considered but the client recognised that these technologies imposed limitations to future land re-sale and/or development and high annual maintenance charges offsetting the predicated savings in cheaper electricity.

A key stage for the Vilter/Star/Cool Partners (VSC) team was the establishment of bespoke selection software as it was essential to know how to assess the overall effectiveness of the system. Previous projects had been assessed in terms of COP. This project needed to be COPc for 'cooling effect/absorbed power'. In addition, COPh for 'heating effect/ divided by absorbed power' was required but the real measure of the heating efficiency is COPhi. This is the total heating capacity divided by the net difference in compressor absorbed power between the design cooling only condensing condition and the heat pump design condensing condition.

Based on the client's previously measured heating and cooling load profiles the analysis showed that to meet the projected hot water heating demands from the 'Total Loss' and 'Closed Loop' circuits, the selected heat pump compressors would have to produce 1.25 MW of high grade heat. To achieve this demand the equipment selected offers 914 kW of refrigeration capacity with an absorbed power rating

of 346 kW. The combined heating and cooling COP, COPhc, is calculated to be a modest 6.25. For an uplift of 17 k in discharge pressure
the increase in absorbed power was 108 kW boosting the COPhi to an
impressive 11.57.

Without the company's commitment to research and logging site
energy performance it would have been impossible to gauge how much
could be saved by using the waste heat from the cooling process. Since
plant handover in May 2010 the company is heating around 54 000 litres
of water each day to 60°C, which costs around £10, saving some
£30 000 pa on gas. By late 2010, the site will be utilising a further
250 kW of waste heat for its Closed Loop systems and by mid-2011, the
demand on all heating systems is set to double.

The Company can save an estimated £143 000 pa in heating costs,
and 119 100 kg in carbon emissions by using a Star Neatpump. Despite
the new refrigeration plant providing both heating and cooling, it consumes £120 000 less electricity per annum than the previous cooling
only plant.

Reproduced by permission of Star Refrigeration Ltd

Chapter 8

Design of Heat Exchangers

Learning outcomes

• Demonstrate the various types of heat exchanger	Knowledge and understanding
• Derive the log mean temperature for heat exchangers	Analysis
• Describe the NTU method for performance of heat exchangers	Problem solving
• Investigate the use of extended surfaces/fins in heat exchangers	Reflections
• Determine fin efficiency and distinguish between the analytical and graphical methods	Analysis
• Estimate the performance of typical heat exchangers	Analysis
• Solve examples concerning the temperature distribution and heat transfer rates in heat exchangers	Problem solving
• Practise further tutorial problems	Problem solving

Energy Audits: A Workbook for Energy Management in Buildings, First Edition.
Tarik Al-Shemmeri.
© 2011 Blackwell Publishing Ltd. Published 2011 by Blackwell Publishing Ltd.

8.1 Types of heat exchanger

Heat transfer has been discussed extensively earlier in this book and in this part we will restrict ourselves to the application of the principles of heat transfer in devices specifically designed to enhance the exchange of heat between two media.

There are two aspects associated with heat transfer. The heat required to raise the temperature of a fluid substance (gas or liquid) is given by:

Heat requirement = mass × specific heat × temperature difference

$$Q = m \ Cp \ \Delta T \qquad\qquad [8.1]$$

The heat exchange from a hot solid surface to the surroundings is given by:

Heat exchange = area × overall heat transfer coefficient × temperature difference

$$Q = A \ U \ \Delta T \qquad\qquad [8.2]$$

A heat exchanger is a device used for transferring heat from a hot fluid to a cold fluid. There are three different types of heat exchanger, depending on the geometry and the way in which the two fluids interact:

● double-pipe heat exchangers;
● shell-and-tube heat exchangers;
● cross-flow heat exchangers.

8.1.1 Double-pipe heat exchangers

The simplest type of heat exchanger, these consist of simple concentric tubes to separate the two fluids. They have two different arrangements (Figure 8.1):

● Parallel-flow. In this type, both fluids flow in the same direction.
● Counter-flow. In this type, the two fluids flow in opposite directions.

8.1.2 Shell-and-tube heat exchangers

This type of heat exchanger is generally bigger in size than its double-pipe counterpart. It has an outer shell containing one fluid, while the other fluid flows inside a number of tubes running in parallel to each other. The shell may have baffles, and the tubes may have fins on the external and even internal surfaces to improve the heat transfer between the two fluids (Figure 8.2).

8.1.3 Cross-flow heat exchangers

In terms of geometry, this is the most complicated of all heat exchangers. In cross-flow heat exchangers, the two fluids flow at right angles to each other.

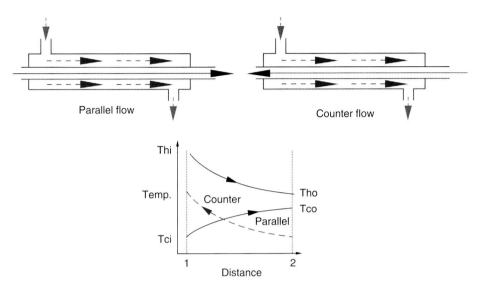

Figure 8.1 Types of double-pipe heat exchanger. Dark grey arrows represent heat; pale grey arrows represent cold.

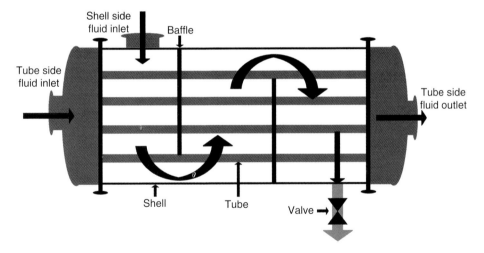

Figure 8.2 Shell-and-tube heat exchanger.

Cross-flow heat exchangers may be further subdivided into:

- Unmixed flow arrangements, where both fluids are confined to travelling through individual passageways within the heat exchanger. In the left-hand diagram of Figure 8.3, both tubes and the cross flow through plates are unmixed.
- Mixed flow arrangements, where one fluid is allowed to travel unimpeded through the heat exchanger enclosure. In the right-hand diagram in Figure 8.3, the tubes are unmixed and the cross flow is mixed.

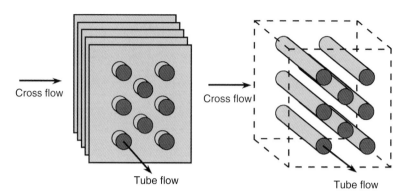

Figure 8.3 Cross-flow heat exchangers.

8.2 Overall heat transfer coefficient

In practical situations, there is a real problem for heat transfer surfaces stemming from the following:

- the build-up of deposits due to a harsh working environment;
- impurities in the working fluid/s;
- change/deterioration of surface condition.

These will add a thermal insulating layer on either side of the heat exchanger surfaces, R_{fi} and R_{fo} respectively (see Table 8.1 for typical values). Hence, the overall heat transfer coefficient (some typical values are given in Table 8.2) will be lowered.

For simplicity, let us consider a typical heat exchanger of tubular configuration. The value of the overall heat transfer coefficient will vary depending on whether it is derived with reference to the tube inner (r_i) or outer radius (r_o), as shown earlier.

With reference to the tube inner radius (r_i):

$$U_i = \cfrac{1}{\cfrac{1}{h_i} + R_{fi} + \cfrac{r_i}{k}\ln\left(\cfrac{r_o}{r_i}\right) + \cfrac{r_i}{r_o}R_{fo} + \cfrac{r_i}{r_o h_o}} \qquad [8.3]$$

With reference to the tube outer radius (r_o):

$$U_o = \cfrac{1}{\cfrac{r_o}{r_i h_i} + \cfrac{r_o}{r_i}R_{fi} + \cfrac{r_o}{k}\ln\left(\cfrac{r_o}{r_i}\right) + R_{fo} + \cfrac{1}{h_o}} \qquad [8.4]$$

The choice is not usually critical, since

$$U_i A_i = U_o A_o \qquad [8.5]$$

Table 8.1 Typical fouling factors.

Type of fluid	Fouling factor, R_f m²K/W
Seawater, below 50°C	0.0001
Seawater, above 50°C	0.0020
Oil	0.0007
Steam, non-oil-bearing	0.0001
Industrial air	0.0004
Refrigerating liquid	0.0002

Table 8.2 Typical overall heat transfer coefficients.

Application	U value W/m²K
Steam condenser	1000–5600
Water-to-water heat exchanger	850–1700
Finned-tube heat exchanger, water in tubes, air across tubes	25–55
Water-to-oil heat exchanger	100–350
Finned-tube heat exchanger, steam in tubes, air over tubes	28–280
Ammonia condenser, water in tubes	850–1400
Gas-to-gas heat exchanger	10–40

8.3 Analysis of heat exchangers

8.3.1 The logarithmic mean temperature difference method

The heat transfer process in a heat exchanger can be divided into three stages: heat lost from hot fluid to the heat exchanger surface, heat transfer from side to side of the heat exchanger surface and, finally, heat lost from the other solid surface to the cold fluid. All three have the same heat source, hence:

$$Q = \dot{m}_h \, Cp_h \, (T_{hi} - T_{ho})$$
$$= U \, A \, \Delta T_m$$
$$= \dot{m}_c \, Cp_c \, (T_{co} - T_{ci}) \qquad [8.6]$$

Here, the subscripts h and c represent hot and cold fluids respectively and i and o stand for inlet and outlet respectively.

Since the fluid inlet/outlet temperatures are usually set by external require-ments, calculation of the rate of heat transfer will depend on the selection of a suitable fluid-to-fluid mean temperature difference (ΔT_m). The generated fluid temperature profiles for counter-flow and co-current flow exchangers are shown below.

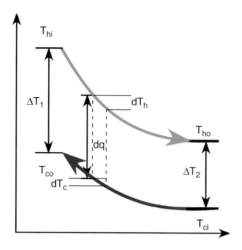

Figure 8.4 Derivation of the log mean temperature difference (LMTD).

Consider the heat transfer dQ across any short section (x–x) of heat exchanger tube (Figure 8.4). The heat transfer from a small section of area dA is:

$$dQ = U\, dA\, \Delta T \tag{8.7}$$

Define the temperature difference at any section by the overall temperature difference:

$$\frac{d(\Delta T)}{\Delta T} = \frac{dT_h - dT_c}{(Q/U)dA} \tag{8.8}$$

Integrating between inlet and outlet gives:

$$\text{Ln}\left(\frac{\Delta T_o}{\Delta T_i}\right) = \frac{\Delta T_o - \Delta T_i}{(Q/U)A} \tag{8.9}$$

Hence

$$Q = UA\left(\frac{\Delta T_o - \Delta T_i}{\text{Ln}\left(\dfrac{\Delta T_o}{\Delta T_i}\right)}\right) \tag{8.10}$$

where

$$\Delta T_m = \frac{\Delta T_o - \Delta T_i}{\text{Ln}\left(\dfrac{\Delta T_o}{\Delta T_i}\right)} = \frac{\Delta T_i - \Delta T_o}{\text{Ln}\left(\dfrac{\Delta T_i}{\Delta T_o}\right)} \tag{8.11}$$

Equation [8.11] is known as the *logarithmic mean temperature difference* (*LMTD*); the two expressions are identical. It can easily be shown that this derivation holds for both parallel and counter-flow arrangements.

8.3.2 The F-method for analysis of heat exchangers

In most practical designs the two fluids will not flow in pure co-current, counter-flow or cross-flow fashion but will be some combination of all three.

In the common shell-and-tube heat exchanger, the temperature profile is further complicated by the fact that the shell-side flow is not in one direction due to the presence of baffles. Baffles are installed to increase shell-side fluid velocities and mixing, and hence improve the shell-side heat transfer coefficient.

Clearly, the simple logarithmic mean temperature difference equation cannot be applied directly in these cases and a correction factor (F) has to be applied to ΔT_m (LMTD) for a simple double-pipe heat exchanger.

The analysis of multipass and cross-flow geometries is usually presented graphically utilising two system characteristic temperature ratios, Z and P. Z is defined below for convenience, but P is known as the heat exchanger effectiveness.

$$Z = \frac{T_i - T_o}{t_o - t_i}, \quad P = \frac{t_o - t_i}{T_i - t_i} \tag{8.12}$$

where T and t denote the shell-side and tube-side temperatures respectively; i and o denote inlet and outlet respectively.

The rate of heat transfer (Q) is given by:

$$Q = U \times A \times \Delta T_m \times F \tag{8.13}$$

There are some correlations for the correction factor, F, which have been developed for shell-and-tube heat exchangers:

$$F = \frac{\left(\dfrac{\sqrt{R^2 + 1}}{R - 1}\right) \ln\left\{\dfrac{1 - X}{1 - RX}\right\}}{\ln\left\{\dfrac{(2/X) - 1 - R + \sqrt{R^2 + 1}}{(2/X) - 1 - R - \sqrt{R^2 + 1}}\right\}}$$

with X defined as:

$$X = \frac{1 - \left[\dfrac{RP - 1}{P - 1}\right]^{1/N}}{R - \left[\dfrac{RP - 1}{P - 1}\right]^{1/N}}$$

where N is the number of shells and P and R ($= Z$) are defined in Equation [8.12]. Correction factor charts for common geometries are presented in Figure 8.5.

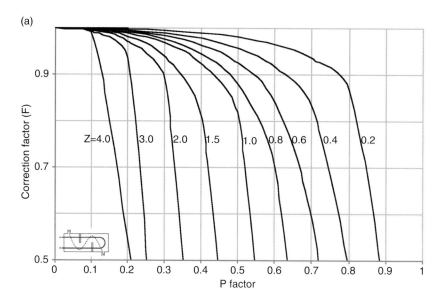

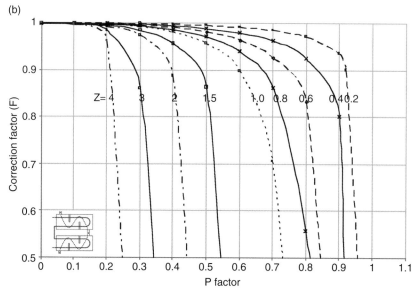

Figure 8.5 (a) Correction factors for shell-and-tube heat exchangers (1 shell); (b) correction factors for shell-and-tube heat exchangers (2 shells);

8.3.3 The effectiveness–NTU method for analysis of heat exchangers

This method for evaluating the performance of heat exchangers has the advantage that it does not require the calculation of the logarithmic mean temperature difference. Its main use is to calculate achievable outlet temperatures by an existing heat exchanger of known area and construction upon a change in operating requirement or duty. The method depends on the evaluation of three dimensionless parameters: the effectiveness, the NTU and the capacity ratio; these are defined in the following paragraphs.

(c)

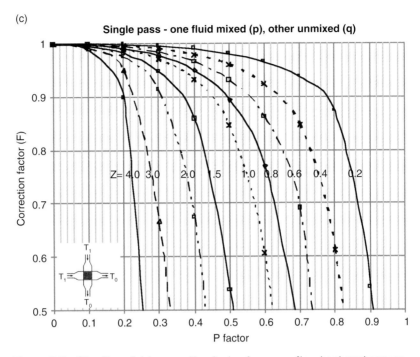

Single pass - one fluid mixed (p), other unmixed (q)

Figure 8.5 *(Continued)* (c) correction factor for cross-flow heat exchangers, one fluid unmixed.

Effectiveness (ε) is defined as the ratio of actual heat exchange to the maximum possible heat transfer, hence:

$$\varepsilon = \frac{\text{Actual heat transfer}}{\text{Maximum possible heat transfer}} = \frac{Q}{Q_{max}} \qquad [8.14]$$

where:

$$Q_{max} = (\dot{m}Cp)_{min}(T_{hi} - T_{ci}) \qquad [8.15]$$
$$Q = \dot{m}_h\, Cp_h\, (T_{hi} - T_{ho}) \qquad [8.16]$$
$$Q = \dot{m}_c\, Cp_c\, (T_{co} - T_{ci})$$

Hence,

$$\varepsilon = \frac{\dot{m}_h Cp_h(T_{hi} - T_{ho})}{(\dot{m}Cp)_{min}(T_{hi} - T_{ci})} \quad \text{or} \quad \frac{\dot{m}_c Cp_c(T_{co} - T_{ci})}{(\dot{m}Cp)_{min}(T_{hi} - T_{ci})} \qquad [8.17]$$

Depending on which data are available.

The number of transfer units (NTU) is defined as:

$$NTU = UA/(\dot{m}\,Cp)_{min} \qquad [8.18]$$

The capacity ratio is defined as:

$$C = (\dot{m}\, Cp)_{min}/(\dot{m}\, Cp)_{max} \qquad\qquad [8.19]$$

It can be shown that, for any heat exchanger, the effectiveness is a function of the NTU, the capacity ratio and the system geometry:

$$\varepsilon = f\,(\text{NTU, capacity ratio, system geometry})$$

The analytical solutions to the above for various geometries are represented graphically in Figure 8.6.

It is possible to use this method in three different ways with any two unknowns, the third can be predicted:

• knowing the effectiveness and the NTU to predict the capacity ratio;
• knowing the effectiveness and the capacity ratio to predict the NTU;
• knowing the NTU and the capacity ratio to predict effectiveness.

When NTU is placed into the effectiveness equations and they are plotted, you can construct the plots shown in Figure 8.6, which are more often used than the equations.

Then, by calculating the C_{min}/C_{max} and the NTU, the effectiveness can be read from these charts. Once the effectiveness has been found, the heat load is calculated by:

$$Q = \text{Effectiveness} \times C_{min} \times (\text{Hot temperature in} - \text{Cold temperature in})$$

Empirical equations for the Effectiveness-NTU Method for double-pipe heat exchangers are as follows:

For coaxial pipes in co-current flow:

$$\varepsilon = \frac{1 - e^{-NTU(1+c)}}{1+c} \qquad\qquad [8.20]$$

For coaxial pipes in counter-current flow:

$$\varepsilon = \frac{1 - e^{-NTU(1+c)}}{1 - ce^{-NTU(1-c)}} \qquad\qquad [8.21]$$

For shell-and-tube – one shell pass and any even tube passes:

$$\varepsilon = \frac{2}{1 + c + \sqrt{1+c^2}\,\dfrac{1 + e^{-NTU\sqrt{1+c^2}}}{1 - e^{-NTU\sqrt{1+c^2}}}} \qquad\qquad [8.22]$$

(a)

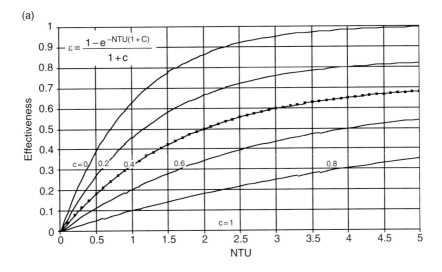

(b)

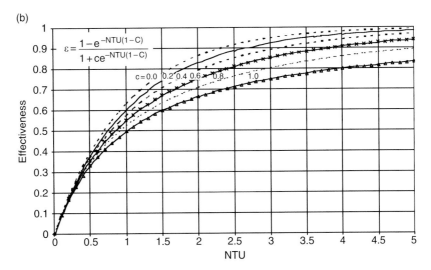

Figure 8.6 (a) NTU–effectiveness chart for parallel-flow heat exchangers; (b) NTU–effectiveness chart for counter-flow heat exchangers;

(c)

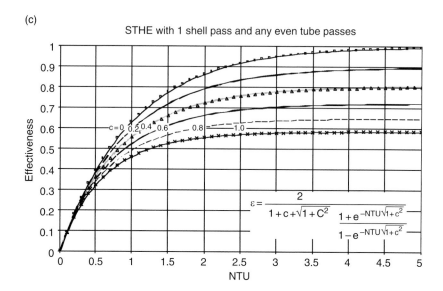

(d)

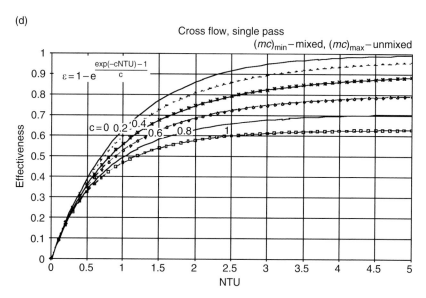

Figure 8.6 (*Continued*) (c) NTU–effectiveness chart for shell-and-tube heat exchangers; (d) NTU–effectiveness chart for cross-flow heat exchangers;

(e)

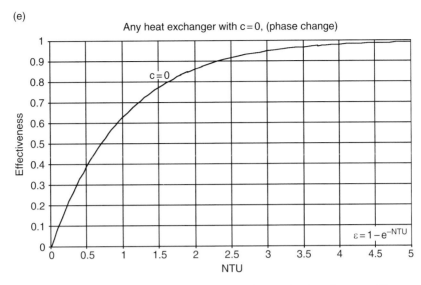

Figure 8.6 (*Continued*) (e) NTU–effectiveness for any heat exchanger with fluid having a phase change.

8.4 Optimisation of heat transfer surfaces (fins)

Convection from a solid surface to the surrounding fluid is limited by the area of the surface. The surface area may be increased by the addition of fins, heat being conducted along the fins and convected to the surrounding fluid. It is likely that the convection coefficient will be altered by the fins, due to the change in flow pattern, and it is probable that it will vary along the fin. Although the surface area has increased, the *average* surface temperature has decreased, due to the exponential temperature decay along the fin. Despite this, the total heat transfer is usually, though not always, increased.

With both finning and surface roughening (in duct flow), the heat transfer rate is increased at the expense of 'pumping power' and the above advantages and disadvantages must be weighed before the degree of finning or surface roughening is decided upon.

8.4.1 Fin types

Fins are usually used on the gas side of heat exchangers in order to compensate for the poor convection coefficients by increasing the heat exchange area. Typical fins are annular fins, which may be of disc or spiral form, plain or corrugated (crimped). Fins may be further classified by the method of attachment to the parent tube. They may be:

- Integral with the base, i.e. rolled on, or extruded, as in aluminium.
- Welded/soldered to the surface - the former are usually resistance welded (ferrous and high nickel only) while the latter are usually attached by running the solder into the joint. Both types are wound on under tension.
- Mechanically attached. This group may be subdivided into:
  - embedded in groove;
  - tension wound - sometimes with inner corrugated edge;
  - tension wound and of L-shaped section.

Thermal contact can be improved in integral and welded/soldered types by soldering, hot-drip galvanising or tinning. Mechanically attached fins have an operating temperature limit due to differential expansion.

8.4.2 Theory of fins

The total heat associated with fins (A_f) and bear surfaces (A_b) is the sum of both:

$$Q_t = Q_f + Q_b$$
$$= h\, A_f\, \eta_F\, \Delta T_b + h\, A_b\, \Delta T_b \qquad\qquad\qquad [8.23]$$

The surface area of different fins (refer to Figure 8.7) is given as:

Rectangular fins $\qquad A_f = 2\,bL$

Triangular fins $\qquad A_f = 2b\left[L^2 + \left(\dfrac{t}{2}\right)^2\right]^{1/2}$

Parabolic fins $\qquad A_f = 2.05\,b\left[L^2 + \left(\dfrac{t}{2}\right)^2\right]^{1/2}$

Annular fins $\qquad A_f = 2\pi\,(r_{2c}^2 - r_1^2)$

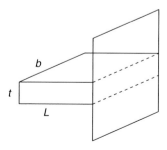

Figure 8.7 Fin dimensions.

The fin efficiency is defined as:

$$\eta_F = \frac{\tanh(m\ell)}{m\ell}$$

[8.24]

where

$$m = \sqrt{\frac{hP_x}{kA_x}}$$

[8.25]

ℓ = effective length of fin = $L + t/2$
P_x = perimeter of fin = $2(b + t)$
A_x = cross-sectional area of fin = $b \times t$
k = thermal conductivity of fin material
h = the convective heat transfer coefficient.

Since fin efficiency is a tedious calculation, there are charts available for typical fin types, and these are used in design calculations (see Figures 8.8 and 8.9).
When the base temperature is not known, i.e. when the temperature difference between the two fluids is specified, the general equation for total heat transfer is given by:

$$\frac{Q}{A_i} = \frac{\Delta T}{\dfrac{1}{\eta_i h_i}\dfrac{A_i}{A_f} + \dfrac{\Delta x}{k} + \dfrac{1}{\eta_o h_o}\dfrac{A_i}{A_f}}$$

[8.26]

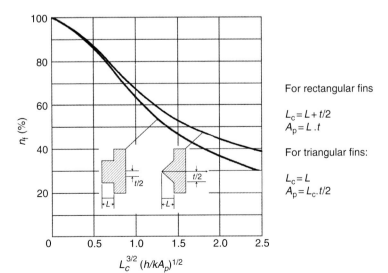

For rectangular fins

$L_c = L + t/2$
$A_p = L . t$

For triangular fins:

$L_c = L$
$A_p = L_c . t/2$

Figure 8.8 Efficiency of straight fins.

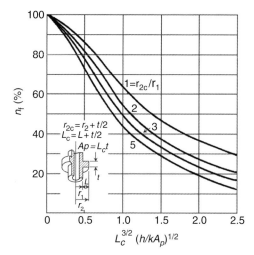

Figure 8.9 Efficiency of annular fins of rectangular profile.

8.5 Worked examples

Worked example 8.1

A concentric heat exchanger has convection heat transfer coefficient values given as $h_i = 2725$ W/m²K, and $h_o = 2100$ W/m²K. If the tube is mild steel, OD = 3.00 cm and ID = 2.50 cm and $k = 45.0$ W/mK, determine the U_o for the following situations:

(a) Normal service (no fouling).
(b) Omitting the conductive resistance of the tube in (a).
(c) If the shell-side fluid is seawater and the fluid inside the tube is treated boiler feedwater. Take fouling factors as: $R_{fi} = 0.0002$ m²K/W and $R_{fo} = 0.00009$ m²K/W.

Solution:

$$r_o = \frac{3.00}{2} = 1.50 \text{ cm} = 0.015 \text{ m} \quad \text{and} \quad r_i = \frac{2.50}{2} = 1.25 \text{ cm} = 0.0125 \text{ m}$$

(a) $U_o = \dfrac{1}{\dfrac{r_o}{r_i h_i} + \dfrac{r_o}{k}\ln\dfrac{r_o}{r_i} + \dfrac{1}{h_o}}$

$$U_o \frac{1}{\dfrac{0.0150}{0.0125 \times 2725} + \dfrac{0.0150 \times \ln(0.0150/0.0125)}{45.0} + \dfrac{1}{2100}}$$

$$= 1023 \text{ W/m}^2\text{K}$$

(b) $U_o \approx \dfrac{1}{0.000440 + 0.000476} = 1091\,\text{W/m}^2\text{K}$

The error in omitting the conductive resistance is:

$= \dfrac{1023 - 1091}{1023} = -6.6\%$

(c) Including fouling effects on the inside and outside surfaces:

$U_o = \dfrac{1}{0.000440 + (1.5/1.25)(0.0002) + 0.0000608 + 0.00009 + 0.000476}$

$= 765\,\text{W/m}^2\text{K}$

This value is 25% lower than the 'clean' value.

Worked example 8.2

We wish to recover the energy from 0.6 kg/s of oil at a temperature of 125°C and cool it down to 65°C by heating water at the rate of 0.5 kg/s at a temperature of 10°C. The overall coefficient of heat transfer is 100 W/m²K. Determine the maximum water outlet temperature and the length of 4-cm ID tubing required for this task. Investigate two possible designs: counter-flow and parallel-flow heat exchange. Use the following data: $Cp_{oil} = 2.10\,\text{kJ/kgK}$; $Cp_{water} = 4.18\,\text{kJ/kgK}$

Solution:

Equating the heat contents of water and oil:

$(\dot{m}C_p\,\Delta T)_w = (\dot{m}\,C_p\,\Delta T)_{oil}$ $(\Delta T)_w = \dfrac{(0.60)(2.10)(60)}{(0.50)(4.18)} = 36.17°C$

Hence

$T_{wo} = 36.17 + 10 = 46.17°C$

For counter flow:

$\Delta T_o = 65 - 10 = 55°C$ and $\Delta T_i = 125 - 46.17 = 78.8°C$

$\text{LMTD} = \dfrac{\Delta T_i - \Delta T_o}{\ln(\Delta T_i/\Delta T_o)} = \dfrac{78.8 - 55}{\ln(78.8/55)} = 66.19°C$

$Q = (\dot{m}\,C_p\,\Delta T)_{oil} = (0.60)(2.10)(60) = 75.6\,\text{kW}$

$A = \dfrac{\Delta Q}{U \times LMTD} = \dfrac{75600}{100 \times 66.19} = 11.421\,\text{m}^2$

$$A = \pi.d.L \qquad L = \frac{11.421}{\pi(0.04)} = 90.9m$$

For parallel flow:

$$\Delta T_i = 125 - 10 = 115°C \quad \text{and} \quad \Delta T_o = 65 - 46.17 = 18.83°C$$

$$LMTD = \frac{\Delta T_i - \Delta T_o}{\ln(\Delta T_i / \Delta T_o)} = \frac{115 - 18.83}{\ln(115 / 18.83)} = 53.14°C$$

$$A = \frac{Q}{U \times LMTD} = \frac{75\,600}{(100) \times (53.14)} = 14.226m^2$$

$$L = \frac{A}{\pi d} = \frac{14.226}{(\pi)(0.04)} = 113.2m$$

The parallel-flow configuration requires a larger area to achieve the same heat transfer as can be achieved by counter flow; hence a counter-flow arrangement is favoured when using double-pipe heat exchangers.

Worked example 8.3

A shell-and-tube oil cooler has the specification outlined in Worked example 8.3, Table 1.

Worked example 8.3, Table 1

Item Fluid	Tube Water	Shell Oil
Diameter (m)	0.04	0.1
Length (m)	1	1
Number	10 tubes × 2 passes	1
Flow rate (kg/s)	3.00	2.67
Specific heat capacity (KJ/kgK)	4.18	2.35
Inlet temperature °C	15	175
U value (W/m2K)	350	

Determine:

(a) The effectiveness of the heat exchanger.
(b) The outlet temperature of oil.

Solution:

(a) For oil $\dot{m}_o\,Cp_o = 2.67 \times 2350 = 6274\,W/°C$
For water $\dot{m}_w\,Cp_w = 3 \times 4180 = 12\,540\,W/°C$

$$C = \frac{(\dot{m}Cp)_{min}}{(\dot{m}Cp)_{max}} = \frac{6274}{12540} = 0.5$$

A = number of tubes × number of passes × surface area of one tube
$\quad = 10 \times 2 \times (\pi DL) = 20 \times (\pi \times 0.04 \times 10) = 25.13\,m^2$

$$NTU = \frac{AU}{(\dot{m}Cp)_{min}} = \frac{25 \times 350}{6274} = 1.39$$

    From the effectiveness chart for shell-and-tube heat exchangers with one shell pass and two tube passes at the capacity ratio and NTU values calculated above, we have effectiveness $(\varepsilon) = 0.63$.

(b) $Q_{max} = (m_o\,Cp_o)\,(T_{oi} - T_{wi}) = 6274 \times (175 - 15) = 1003\,kW$

Using the effectiveness definition for oil, $\varepsilon = Q_{oil}/Q_{max}$

$$T_{o2} = T_{o1} - \frac{\varepsilon \times Q_{max}}{(mCp)_{oil}} = 175 - \frac{0.63 \times 1003 \times 10^3}{6274} = 74.2°C$$

Worked example 8.4

A single-shell-and-two-tube-pass heat exchanger is used to recover energy from wastewater at 30°C to heat fresh water entering at 15°C. The mass flow rate of the wastewater is 2 kg/s and that of the fresh water is 1 kg/s. Using the data given below, calculate:

(a) The effectiveness of the heat exchanger.
(b) The temperature of the fresh water at exit.
(c) Tube length and diameter if the velocity of flow is to be limited to 0.3 m/s and the pass length to be restricted to 2 m.

Use the following data:

    Total heat transfer area = 10 m²

Specific heat of wastewater and fresh water = 4200 J/kgK
Density of water = 1000 kg/m³
Overall heat transfer coefficient = 1250 W/mK
For a single-shell pass, two-tube pass heat exchanger, take the following characteristic as valid:

$$E = \frac{1 - e^{-NTU(1-C)}}{1 - C \times e^{-NTU(1-C)}} \qquad \text{where } C = Cp_{min}/Cp_{max}$$

Solution:

(a) $(\dot{m} C_p)_h = 2 \times 4200 = 8400$ and $(\dot{m} C_p)c = 1 \times 4200 = 4200$

Hence

$$C = \frac{(mCp)_{min}}{(mCp)_{max}} = \frac{4200}{8400} = 0.5$$

$$NTU = \frac{A.U}{(mCp)_{min}} = \frac{10 \times 1250}{4200} = 3$$

$$E = \frac{1 - e^{-NTU(1-C)}}{1 - C \times e^{-NTU(1-C)}} = \frac{1 - e^{-3(1-0.5)}}{1 - 0.5 \times e^{-3(1-0.5)}} = 0.87$$

(b) $Q_{max} = (\dot{m} C_p)_{min} \times (\Delta T)_{max} = 4200 \times (30 - 15) = 63\,000 \text{ W}$

$Q_h = E \times Q_{max} = (\dot{m} C_p)_h \times (\Delta T)_h$

$\qquad = 0.87 \times 63\,000 = 8400 \times (30 - T_{h2})$

Hence

$T_{h2} = 23.48°C$

(c) $\dot{m} = \rho AV$

So

$2 = 1000 \times A_t \times 0.3$

Hence

$A_t = 0.00667 \text{ m}^2$

$$A_t = \frac{\pi}{4} D^2$$

$$\therefore \quad D = \sqrt{\frac{4A_L}{\pi}} = 0.092\text{m}$$

Therefore, we should use 10 mm.

Since

$A_L = 10\,\text{m}^2 = \pi DL \times n$

2-tube pass, total length $L = 2 \times 2 = 4\,\text{m}$

$$\therefore \quad n = \frac{10}{\pi \times 0.010 \times 4} = 79.6$$

Therefore, 80 tubes will be required.

Worked example 8.5

In a chemical plant, a liquid (density 1100 kg/m³, specific heat 4.60 kJ/kgK) is to be heated from 65°C to 100°C by passing it through a tubular heat exchanger where it will be heated by wet steam at 115°C. The liquid flow rate required is 12.96 kg/s and, for design purposes, an average liquid velocity of 1.2 m/s has been chosen. The tubes carrying the liquid are to be stainless steel, 25 mm bore, 1.5 mm thick, and their length must not exceed 3.5 m. Values for the tubes' inside and outside heat transfer coefficient may be taken as 5 kW/m²K and 10 kW/m²K respectively, and the thermal conductivity of steel is 40 W/mK.

(a) Calculate the logarithmic mean temperature difference for the heat exchanger.
(b) Estimate the number of parallel tubes required to cater for the liquid flow.
(c) Determine the number of tube passes needed.

Solution:

(a) The log mean temperature:

$$\Delta T_m = \frac{\Delta T_i - \Delta T_o}{\ln(\Delta T_i / \Delta T_o)}$$

$$= \frac{(115 - 65) - (115 - 100)}{\ln(115 - 65) / (115 - 100)}$$

$$= 29.07\,\text{K}$$

A plot is shown in Worked example 8.5, Figure 1.

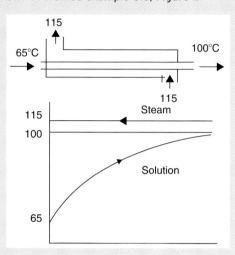

Worked example 8.5, Figure 1

(b) Let N = number of parallel tubes

$$\dot{m} = \rho A V \times N$$

$$12.96 = 1100 \cdot \frac{\pi}{4} \cdot 0.025^2 \times 1.2 \times N$$

$$N = 20 \text{ tubes}$$

(c) For heat gained by liquid:

$$\dot{Q} = \dot{m} \, \text{Cp} \, (T_o - T_i) = 11.8 \times 4.60 \, (100 - 65) = 1900 \, \text{kW}$$

$$U_i = \frac{1}{\frac{1}{h_i} + \frac{r_i}{k} \ln\left(\frac{r_o}{r_i}\right) + \frac{r_i}{r_o h_o}} = \frac{1}{\frac{1}{5000} + \frac{0.0125}{400} \times \ln\frac{0.014}{0.0125} + \frac{0.0125}{0.014 \times 10000}}$$

$$= 3079 \, \text{W/m}^2\text{K}$$

$$\dot{Q} = U \cdot A \cdot \Delta T_m$$

$$1900 = 3.079 \times A \times 29.07$$

$$\therefore A = 21.222 \, \text{m}^2$$

Area per tube = 21.222/20 = 1.061 m²

$$A = \pi D L$$

Hence

$$L = 1.061/ \, (\pi \times 0.025) = 13.5 \, \text{m}$$

Number of passes $= \dfrac{\text{Total length}}{\text{Length per pass}} = \dfrac{13.5}{3.5} = 3.86$

Therefore we need four passes.

Worked example 8.6

Compare the fin efficiencies for four materials (copper, aluminium, stainless steel and glass) if the fin is straight rectangular with a thickness (t) of 20 mm, a length (L) of 100 mm and a width (b) of 1 m in an environment where the convection coefficient is 25 W/m²K.

Solution:

$P = 2b + 2t = 2 (1.00 + 0.02) = 2.04\,\text{m}$
$A = b \times t = 1 \times 20 \times 10^{-3} = 0.020\,\text{m}^2$
$L_c = L + t/2 = 100 + 20/2 = 0.11\,\text{m}$

$$m = \sqrt{\frac{hP_x}{kA_x}} = \sqrt{\frac{25 \times 2.04}{0.02}} \sqrt{\frac{1}{k}} = \frac{50.5}{\sqrt{k}}$$

The relevant data are summarised in Worked example 8.6, Table 1.

Worked example 8.6, Table 1

Material	k	$m = \sqrt{\dfrac{hP}{kA}}$	$\eta_f = \dfrac{\tanh(mL_c)}{mL_c}$
Copper	385	2.574	0.974
Aluminium	170	3.873	0.943
Stainless steel	17	12.248	0.648
Glass	0.8	56.461	0.161

Graphical solution:

$L_c = L + t/2$
$\quad = 100 + 20/2$
$\quad = 0.11\,\text{m}$

$A_p = L_c \times t = 0.0022\,\text{m}^2$

$$L_c^{1.5} \sqrt{\frac{hP}{kA_p}} = \frac{3.889}{\sqrt{k}}$$

The relevant data are summarised in Worked example 8.6, Table 2.

Worked example 8.6, Table 2

Material	k	$L_c^{1.5}\sqrt{\dfrac{hP}{kA_p}} = \dfrac{3.889}{\sqrt{k}}$	η_f from curve
Copper	385	0.198	0.98
Aluminium	170	0.298	0.94
Stainless steel	17	0.943	0.63
Glass	0.8	4.348	< 0.30

This example demonstrates that both solutions are in agreement.

Worked example 8.7

Investigate the suitability of straight rectangular fins made from iron ($k = 50$ W/mK) with length 100 mm, thickness 20 mm and width 1m for an environment where the convection heat transfer coefficient is in the range: $10 < h < 100$ W/m²K for air and $500 < h < 5000$ W/m²K for water.

Solution:

With reference to Worked example 8.7, Figure 1:

$P = 2b + 2t = 2 (1.00 + 0.02) = 2.04$ m
$A = b \times t = 1 \times 20 \times 10^{-3} = 0.020$ m²
$L_c = L + t/2 = 100 + 20/2 = 0.11$ m

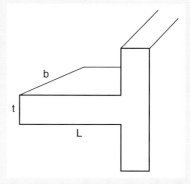

Worked example 8.7, Figure 1

$$m = \sqrt{\frac{hP_x}{kA_x}} = \sqrt{\frac{2.04}{0.02}}\sqrt{\frac{h}{k}} = 10.1\sqrt{\frac{h}{k}}$$

The relevant data are summarised in Worked example 8.7, Table 1.

Worked example 8.7, Table 1

H	M	m_l	η_t
10	5.52	0.497	0.925
100	14.28	1.57	0.584
500	31.94	3.51	0.28
5000	101	11.1	0.09

This example proves that fins are not suitable for liquids.

Worked example 8.8

Investigate the effect of weight on heat transfer for three materials: copper, aluminium and stainless steel in an environment where the convection heat transfer coefficient, $h = 25\,W/m^2K$. Investigate the suitability of rectangular fins of length 100 mm, thickness 20 mm and width 1 m.

Solution:

With reference to Worked example 8.8, Figure 1:

$$P = 2b + 2t = 2\,(1.00 + 0.02) = 2.04\,m$$
$$A = b \times t = 1 \times 20 \times 10^{-3} = 0.020\,m^2$$
$$L_c = L + t/2 = 100 + 20/2 = 0.11\,m$$

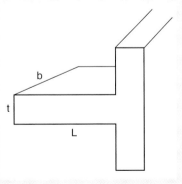

Worked example 8.8, Figure 1

$$m = \sqrt{\frac{hP_x}{kA_x}} = \sqrt{\frac{2.04}{0.02}}\sqrt{\frac{h}{k}} = 10.1\sqrt{\frac{h}{k}}$$

The relevant data are summarised in Worked example 8.8, Table 1.

Worked example 8.8, Table 1

Material	k	ρ	Weight ratio	η_f	Q	Q/W
Copper	385	970	3.464	0.974	1.033	0.290
Aluminium	170	280	1	0.943	1	1
Stainless steel	17	855	3.053	0.648	0.687	0.225

Aluminium is clearly lighter than both other materials, with a weight ratio of nearly 1/3. Although aluminium has a lower fin efficiency than copper, its Q/W is four times better than both other options.

Worked example 8.9

The cylinder barrel of a motorcycle is constructed of aluminium alloy and has height, $H = 0.15\,m$ and outside diameter, $D = 50\,mm$. Under typical operating conditions, the outer surface of the cylinder is at a temperature of 500 K and is exposed to ambient air at 300 K, with a convection coefficient of 50 W/m²K. Annular fins of rectangular profile are added to increase heat transfer to the surroundings. Assume that five such fins ($k = 150$ W/mK), which are of thickness $t = 6\,mm$, length $L = 20\,mm$ and equally spaced, are added. What is the increase in heat transfer due to the addition of the fins?

Solution:

With reference to Worked example 8.9, Figure 1, without the fins, the heat transfer rate is:

$$Q_{wo} = h.A_{wo}\,(T_b - T_\infty)$$

where the area of the unfinned surface is $A_{wo} = H \times 2\pi r_1$

$$Q_{wo} = 50\,(0.15 \times 2\pi \times 0.025)\,(500 - 300) = 236\,W$$

For a single circular fin:

$$P = 2\pi r_2 + 2t = 2\pi \times 0.045 + 2 \times 0.006 = 0.295\,m$$
$$A = 2\pi r_2 \times t = 2\pi \times 0.045 \times 0.006 = 1.696 \times 10^{-3}\,m^2$$
$$m = \sqrt{\frac{hP}{kA}} = \sqrt{\frac{50 \times 0.295}{150 \times 1.696 \times 10^{-3}}} = 7.6$$
$$\eta_f = \frac{\tanh(ml)}{ml} = \frac{\tanh(7.6 \times 0.02)}{7.6 \times 0.02} = 0.99$$

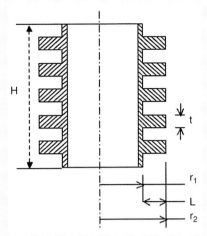

Worked example 8.9, Figure 1

Area of bare tube, less finned surface:

$$A_b = L_b \times 2\pi r_1 = (0.15 - 5 \times 0.006) \times 2\pi \times 0.025 = 0.01885\,m^2$$

Area of finned surface (five fins):

$$A_f = 5\,[2 \times \pi \times (r_2^2 - r_1^2)] = 0.044\,m^2$$

Heat loss from bare tube:

$$Q_b = h\,A_b\,\Delta T_b = 50 \times 0.01885 \times (500 - 300) = 188\,W$$

Heat loss from finned surface only:

$$Q_f = h\,A_f\,\eta_F\,\Delta T_b = 50 \times 0.99 \times 0.044 \times (500 - 300) = 435\,W$$

Total heat transfer:

$$Q_t = Q_f + Q_b = 188 + 435 = 623\,W$$

In other words, fins increased heat transfer by more than three times.

Worked example 8.10

Water flowing in a 4 cm OD tube is to be heated by combustion gases flowing over the tube. Heat transfer to the water is increased with annular fins of plain carbon steel (k = 57 W/mK) attached to the outer surface of the tube. The fins have 4 cm ID, 12 cm OD and are 1 mm thick. The convective heat transfer coefficient on the gas side is 20 W/m²K.

(a) Determine the fin efficiency.

(b) In order to improve the heat transfer rate, two possibilities are being considered:

- increase the outer radius, keeping the thickness at 1mm;
- increase the thickness, keeping the outer radius at 6cm.

Which would you recommend? Show your reasoning by calculation.

Solution:

(a) $L = \dfrac{0.12}{2} - \dfrac{0.04}{2} = 0.04\,\text{m}$

With reference to Worked example 8.10, Figure 1,

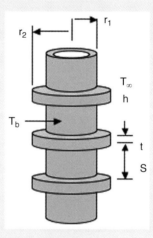

Worked example 8.10, Figure 1

$$m = \sqrt{\dfrac{hP}{kA}} = \sqrt{\dfrac{h \times 2(\pi D_2 + t/2)}{k \times (\pi D_2 \times t)}}$$

$$= \sqrt{\dfrac{20 \times 2(\pi \times 0.12 + 0.001)}{57 \times (\pi \times 0.12 \times 0.001)}} = 41.94$$

and

$$\eta_f = \dfrac{\tanh(ml)}{ml} = \dfrac{\tanh(41.94 \times 0.04)}{41.94 \times 0.04} = 0.55$$

(b) For the first possibility, if r_2 is increased by 1cm while t is kept constant at 1mm:

$$L = \dfrac{0.14}{2} - \dfrac{0.04}{2} = 0.05\,\text{m}$$

$$m = \sqrt{\dfrac{hP}{kA}} = \sqrt{\dfrac{h \times 2(\pi D_2 + t)}{k \times (\pi D_2 \times t)}} = \sqrt{\dfrac{50 \times 2(\pi \times 0.14 + 0.001)}{57 \times (\pi \times 0.14 \times 0.001)}} = 41.9$$

and

$$\eta_f = \dfrac{\tanh(ml)}{ml} = \dfrac{\tanh(41.9 \times 0.05)}{41.9 \times 0.05} = 0.46$$

For the second possibility, if t is increased, say to 2 mm, while the outer radius is kept as in (a):

$$m = \sqrt{\frac{hP}{kA}} = \sqrt{\frac{h \times 2(\pi D_2 + t)}{k \times (\pi D_2 \times t)}} = \sqrt{\frac{50 \times 2(\pi \times 0.12 + 0.002)}{57 \times (\pi \times 0.12 \times 0.002)}} = 26.69$$

and

$$\eta_t = \frac{\tanh(ml)}{ml} = \frac{\tanh(26.69 \times 0.04)}{26.69 \times 0.04} = 0.70$$

It is obvious, therefore, that increasing the thickness is the best option here.

8.6 Tutorial problems

8.1 An energy recovery system consists of a shell-and-tube heat exchanger with four tube passes contained in two shells, the hot stream flowing through the shell side and the cold stream flowing through the tubes. If the operating temperatures are:

Hot fluid inlet = 75°C
Hot fluid outlet = 45°C
Cold fluid inlet = 15°C
Cold fluid outlet = 35°C

estimate the corrected counter-flow LMTD for the heat exchanger and determine the amount of thermal energy recovered from the process. The overall heat transfer coefficient for this heat exchanger is 300 W/m²K, and the effective surface area of the heat exchanger is 10 m².
Ans. (33 K, 99 kW)

8.2 It is desired to cool 0.6 kg/s of oil from 125°C to 65°C. Water is available with a flow rate of 0.5 kg/s at a temperature of 10°C. The overall coefficient of heat transfer is 85 W/m²K. Determine the length of 5-cm ID tubing required for counter-flow and for parallel-flow heat exchange. Use the following data: $Cp_{oil} = 2.10$ kJ/kgK; $Cp_{water} = 4.18$ kJ/kgK.
Ans. (85.5 m, 106.5 m)

8.3 The heat transfer coefficients on the inside and outside of the inner tube in a concentric tube heat exchanger are 200 and 2000 W/m²K respectively. If the tube inner and outer diameters are 20 mm and 23 mm respectively and the thermal conductivity of the

tube is 100 W/mK, calculate the U value for the tube in a clean condition based on the inside area and the outside area.
Ans. (183.5, 159.6 W/m²K)

8.4 A single-shell pass, two-tube pass oil cooler has the specification shown in Tutorial problem 8.4, Table 1.

Tutorial problem 8.4, Table 1

| Item | Tube | Shell |
Fluid	Water	Oil
Flow rate (kg/s)	3	2
Specific heat capacity (kJ/kgK)	4.2	2.1
Inlet temperature °C	15	175

Heat exchanger area = 25 m², U value = 350 W/m²K
Determine the effectiveness and the outlet temperature of the oil.
Ans. (0.75, 54.9°C)

8.5 Water flowing at the rate of 0.1 kg/s is heated from 40°C to 80°C in a counter-flow double-pipe heat exchanger. The hot fluid is oil and the overall heat transfer coefficient is 250 W/m²K. If the oil enters at 105°C, determine the heat transfer area required to cool it to 70°C.
Ans. (2.45 m²)

8.6 Compare the fin efficiencies for stainless steel and aluminium if the fin is straight rectangular with a thickness of 2 mm, a width of 1 m and a length of 100 mm; the convection coefficient is 50 W/m²K.
Ans. (18%, 54%)

8.7 Investigate the suitability of straight rectangular fins made from aluminium (k = 170 W/mK) if the convection heat transfer coefficient is in the range: 10 < h < 100 W/m²K for gas applications and 1000 < h < 10 000 W/m²K for liquid applications.
Ans. (at h = 10, Eff = 84%, at h = 10 000, Eff = 4%)

8.8 Aluminium fins (k = 200 W/mK) of length 15 mm and thickness 1 mm are placed on a 25 mm external diameter tube. The fin base is maintained at 125°C while the ambient temperature is 25°C, with a convection heat transfer coefficient of 100 W/m²K. Determine the heat loss per fin.
Ans. (1.75 kW)

8.9 The cylinder barrel of a motorcycle is constructed of 2024-T6 aluminium alloy and has height, H = 0.15 m and outside diameter, D = 50 mm. Under typical operating conditions, the outer surface of the cylinder is at a temperature of 400 K and is exposed to ambient air at 300 K, with a convection coefficient of 50 W/m²K. Annular fins of rectangular profile are typically added to increase heat transfer to the surroundings. Assume that five such fins, which are of thickness $t = 4$ mm, length $L = 20$ mm and equally spaced, are added. What is the increase in heat transfer due to the addition of the fins?

Ans. (3 times, 621 W)

Chapter 9

Instrumentation for Energy Management

Learning outcomes

• Demonstrate the need for instrumentation in energy management	Knowledge and understanding
• Describe the different methods of measuring temperature	Knowledge and understanding
• Describe the different methods of measuring relative humidity	Knowledge and understanding
• Describe the different methods of measuring pressure and flow rates	Knowledge and understanding
• Describe the different methods of measuring electrical current, voltage and wattage	Knowledge and understanding
• Describe the criteria for selection of sensors and instrumentation	Analysis and reflections
• Solve problems related to instrumentation in energy management	Problem solving
• Practise further tutorial problems	Problem solving

Energy Audits: A Workbook for Energy Management in Buildings, First Edition.
Tarik Al-Shemmeri.
© 2011 Blackwell Publishing Ltd. Published 2011 by Blackwell Publishing Ltd.

9.1 Introduction

The control of energy consumption necessitates the measurement of the following properties:

- temperature;
- humidity;
- pressure;
- flow rate;
- electrical power.

9.2 Temperature measurement

To be considered for use in the measurement of temperature, a substance must experience some recognisable change when its temperature is changed. Moreover, this change must be repeatable without deterioration.

The following material properties are used to measure temperature:

- change in size;
- change in pressure if enclosed;
- change of electrical resistance;
- generation of an electromotive force;
- change of state;
- change of colour;
- change in degree of surface radiation.

The choice of material and effect used for measurement is governed by the temperature range considered, the degree of accuracy required, the type of installation and the original cost. The general grouping of thermometric devices will be considered in the following sections.

9.2.1 Expansion thermometers

This type of thermometer relies on the relationship between the volume and temperature of a solid, liquid or gas.

Liquid-in-glass thermometers

This type of thermometer utilises the expansive properties of a liquid with temperature. The liquid is contained in a very fine glass capillary tube, which terminates in an enlarged part called the bulb of the thermometer (Figure 9.1). Before the tube is sealed off at its upper end, the space above the liquid is exhausted of air and filled with an inert gas such as nitrogen. Historically mercury-in-glass thermometers were used widely, but many countries have now banned them outright due to the toxicity of mercury, which has been replaced by other liquids such as alcohol, acetone, aniline and others. Since these are transparent, the

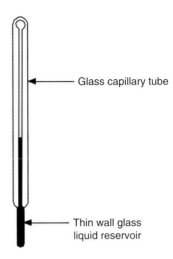

Figure 9.1 Liquid-in-glass thermometer.

Table 9.1 Freezing and boiling temperatures for typical liquids.

Liquid	Freezing temperature/°C	Boiling temperature/°C
Water	0	100
Alcohol	−112	78
Mercury	−39	357

liquid is made more visible by the addition of a red or blue dye. The range of usefulness of the thermometer is set by the boiling point of the liquid used (Table 9.1). In the case of an ethanol-filled thermometer, the upper limit for measurement is 78°C, which makes it useful for measuring body temperature.

Liquid-in-steel thermometers

A liquid-in-steel thermometer consists of a strong steel bulb 150–200 mm long and about 6 mm in diameter, connected by a length of special steel tubing of very fine bore to a coiled tube contained in a circular case. The coiled tube operates a pointer, which moves around a circular dial (Figure 9.2). The whole system is filled with a known liquid, and when the liquid in the bulb expands or contracts, the gauge tube moves out or in and causes the pointer to move an amount equivalent to the temperature rise or fall. This instrument is calibrated by placing the bulb in ice, steam, sulphur vapour and molten silver corresponding to 0°C, 100°C, 444.6°C and 960.8°C respectively.

Bi-metal thermometers

These make use of the difference in the expansion of two different metal strips. They are robust and suitable for relatively high-temperature industrial applications.

Figure 9.2 Dial for a liquid-in-steel thermometer.

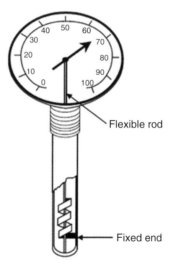

Figure 9.3 Bi-metal thermometer.

The thermometer consists of two strips of dissimilar metal firmly attached to each other throughout their length, as shown in Figure 9.3. The strip is normally enclosed in a metal capsule, and the extension or contraction is monitored by a coil which, in turn, is connected to an arrow rotating on a calibrated scale.

Of interest, this same bi-metal strip idea is used in thermostats for the control of temperature. They are arranged such that, at a given pre-set temperature, they interrupt an electric heater circuit, thus cutting off the power (e.g. in electric kettles). When the temperature begins to fall, they eventually cut in the electric circuit again, thus restoring the power.

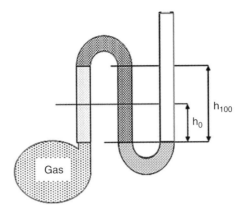

Figure 9.4 Gas thermometer.

Gas thermometers

Gases possess a very high coefficient of thermal expansion about three hundred and sixty times that of glass. Such gases as argon, hydrogen, helium, nitrogen and oxygen can be obtained with a high degree of purity; they can be heated or cooled through very large ranges of temperature without any appreciable deviation from a uniform rate of expansion, and they behave very nearly as perfect gases. For 0 to 210°C, helium is the ideal gas. Hydrogen is used for 0–550°C. Nitrogen comes last and is used for the upper ranges of temperature up to 1650°C.

The variation of temperature with volume is, therefore, governed by Charles's Law (at constant pressure) given by:

$$\frac{V_1}{V_2} = \frac{273 + T_1}{273 + T_2} \qquad [9.1]$$

A gas thermometer is designed such that the variation in the volume of a gas is measured by a U-tube filled with a liquid such that the difference (h) in the level of each side is zero at 0°C. Therefore:

$$T_x = 373 \times \frac{h_x - h}{h_{100} - h_0} \qquad [9.2]$$

The system is usually adjusted so that $h_0 = 0$ (see Figure 9.4).

9.2.2 Electrical resistance thermometers

Only two types of thermometer based on the relationship between temperature and the electrical resistance of the material are considered here: resistance temperature detectors and thermistors.

Resistance temperature detectors (RTDs)

This type of thermometer is based on the fact that the electrical resistance of metal increases as the temperature of the metal is raised.

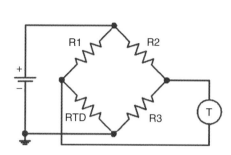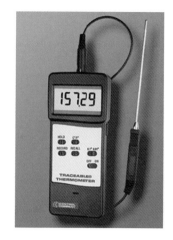

Figure 9.5 Platinum resistance thermometer and its circuit on the left.

A resistance thermometer consists of a coil of wire suitably wound and enclosed in a sheath. The active part of the measuring circuit R_T is encompassed in a Wheatstone bridge (Figure.9.5).

The relationship between resistance and temperature is very nearly linear and follows the equations:

For < 0°C $\qquad R_T = R_0 [1 + aT + bT^2 + cT^3 (T - 100)]$ $\qquad\qquad$ [9.3]

For > 0°C $\qquad R_T = R_0 [1 + aT + bT^2]$ $\qquad\qquad\qquad\qquad$ [9.4]

where

R_T = resistance at temperature T
R_0 = nominal resistance
a, b and c are constants used to scale the RTD.

Platinum resistance thermometers are electrical thermometers which make use of the variation of resistance of high-purity platinum wire with temperature. This variation is predictable, enabling accurate measurements to be performed to better than a thousandth part of 1°C. The resistance/temperature curve for a typical platinum RTD is shown in Figure 9.6.

Thermistors

Thermistors have a semiconductor material which changes its electrical resistance as a function of temperature (Figure 9.7) and are manufactured from the oxides of the transition metals - manganese, cobalt, copper and nickel. The typical thermistor has a negative temperature coefficient. This means that

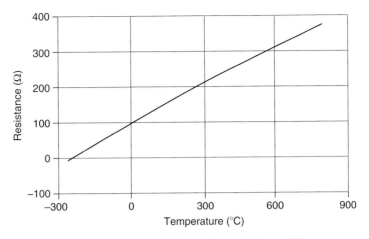

Figure 9.6 Resistance–temperature curve for a platinum RTD.

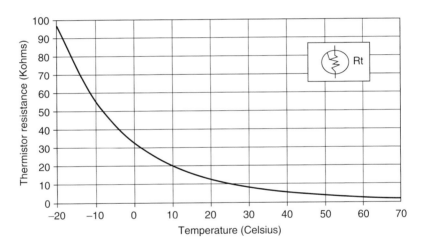

Figure 9.7 Thermistor characteristics for a typical sensor.

with an increase in temperature, the resistance of the thermistor decreases. Although the thermistor is a fairly sensitive device, it is very nonlinear and usually used over a very small temperature span. In addition, thermistors are quite fragile and great care must be taken to mount them so that they are not exposed to shock or vibration.

The standard formula for NTC thermistor resistance as a function of temperature is given by:

$$R = R_0 \times e^{B \times (1/T - 1/T_0)} \qquad\qquad [9.5]$$

where the temperatures are in Kelvin; R_0 is the resistance at temperature T_0 (usually 25°C = 298.15°K) and $B = 10^4/2.37$.

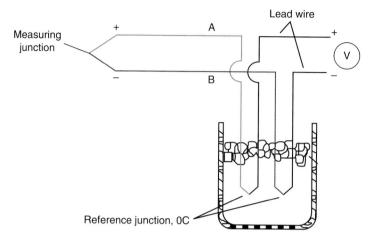

Figure 9.8 Thermocouple. Dark grey lines represent positive polarity; pale grey lines represent negative polarity.

9.2.3 Thermocouples

Thermocouples are based on a discovery made by Thomas Johann Seebeck. In 1821 he found that if two dissimilar materials are joined together at their ends (see Figure 9.8) and one of the junctions is heated, an electrical current will flow in the circuit. The value of the emf developed depends upon the difference in the temperature of the hot and cold junctions. Thus, if one of the junctions is maintained at ice point (0°C), the emf will be a direct measure of the temperature at the hot junction. The reference junction at zero degrees can be substituted by an electronic circuit which is housed in the electronic read-out box. Due to the very small values of emf, the signal is usually magnified because normal voltmeters are designed to read relatively large values (volts rather than microvolts associated with typical thermocouples' output).

Table 9.2 Properties of thermocouple pairs.

Type	Combination	Range	emf mv at 100°C
S	Platinum – 10% Rhodium v Platinum	0–1400°C	0.645
R	Platinum – 13% Rhodium v Platinum	0–1400°C	0.647
J	Iron/Copper-Nickel	0–800°C	5.268
K	Nickel-Chromium/ Nickel-Aluminium	0–1100°C	4.095
T	Copper/Copper-Nickel	−200–400°C	4.277
E	Nickel-Chromium/ Copper Nickel	0–800°C	6.137

Table 9.2 lists some of the most common thermocouple material pairs in use. Each pair is designated with a letter; T and K are the most common types used in engineering applications.

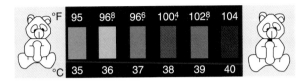

Figure 9.9 Forehead phase change thermometer.

9.2.4 Change-of-state thermometers

These temperature sensors consist of labels, pellets, crayons, lacquers or liquid crystals whose appearance changes once a certain temperature is reached. They are used, for instance, with steam traps – when a trap exceeds a certain temperature, a white dot on a sensor label attached to the trap will turn black. Response time typically takes minutes, so these devices often do not respond to transient temperature changes. Accuracy is also lower than with other types of sensor. Furthermore, the change in state is irreversible, except in the case of liquid-crystal displays. Even so, change-of-state sensors can be handy when one needs confirmation that the temperature of a piece of equipment or a material has not exceeded a certain level, for instance for technical or legal reasons during product shipment. There are medical strips based on this principle operating over a narrow range of body temperatures (34-38°C), as shown in Figure 9.9.

9.2.5 Optical pyrometers

Optical pyrometers (Figure 9.10) work on the basic principle of using the human eye to match the brightness of a hot object to the brightness of a calibrated lamp filament inside the instrument. The optical system contains filters that restrict the wavelength sensitivity of the devices to a narrow wavelength band around 0.65 to 0.66 microns (the red region of the visible spectrum). Other filters reduce the intensity so that one instrument can have a relatively wide temperature range capability. Needless to say, by restricting the wavelength response of the device to the red region of the visible spectrum, it can only be used to measure objects that are hot enough to be incandescent, or glowing. This limits the lower end of the temperature measurement range of these devices to about 700°C. Some experimental devices have been built using light amplifiers to extend the range downwards, but the devices become quite cumbersome, fragile and expensive.

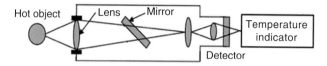

Figure 9.10   Optical pyrometer.

Figure 9.11   Infrared thermometer.

9.2.6 Infrared temperature sensors

These measure the amount of radiation emitted by a surface. Electromagnetic energy radiates from all matter regardless of its temperature. In many process situations, the energy is in the infrared region. As the temperature goes up, the amount of infrared radiation and its average frequency go up.

Different materials radiate at different levels of efficiency. This efficiency is quantified as emissivity, a decimal number or percentage ranging between 0 and 1 or 0% and 100%. Most organic materials, including skin, are very efficient, frequently exhibiting emissivities of 0.95. Most polished metals, on the other hand, tend to be inefficient radiators at room temperature, with emissivity or efficiency often 20% or less.

To function properly, an infrared measurement device (Figure 9.11) must take into account the emissivity of the surface being measured. This can often be looked up in a reference table. However, bear in mind that tables cannot account for localised conditions such as oxidation and surface roughness. Sometimes a practical way to measure temperature with infrared when the emissivity level is not known is to 'force' the emissivity to a known level by covering the surface with masking tape (emissivity of 95%) or a highly emissive paint. An infrared device is like a camera and thus covers a certain field of view. It might, for instance, be able to 'see' a 1-deg visual cone or a 100-deg cone. When measuring a surface, be sure that the surface completely fills the field of view. If the target surface does not, at first, fill the field of view, move closer or use an instrument with a narrower field of view. Or simply take the background temperature into account (i.e., adjust for it) when reading the instrument.

9.2.7 Selection guides for temperature measurement

Choosing the correct temperature measurement device depends on various technical factors as well as cost. Each device has its special features and characteristics, such as application environment, range of temperature, accuracy, physical strength and so on.

RTDs are more stable than thermocouples. On the other hand, as a class, their temperature range is not as broad: RTDs operate from about −250 to 850°C whereas thermocouples range from about −270 to 2300°C. Thermistors have a more restrictive span, being commonly used between −40 and 150°C, but offer high accuracy in that range.

Thermistors and RTDs share a very important limitation. They are resistive devices and accordingly they function by passing a current through a sensor.

Infrared sensors, though relatively expensive, are appropriate when the temperatures are relatively high. They are available for up to 3000°C, far exceeding the range of thermocouples or other contact devices. The infrared approach is also attractive for noncontact applications. Thus, fragile or wet surfaces, such as painted surfaces coming out of a drying oven, can be monitored in this way. Substances that are chemically reactive or electrically noisy are ideal candidates for infrared measurement.

9.3 Humidity measurement

Moisture content and relative humidity are two interlinked terms which are often used in defining the comfort level of buildings and play an important role in the behaviour of air-conditioning systems. In order to define and control such a parameter, it is important to measure it. There are many methods used in the measurement of humidity or moisture content. The relevant instruments are called humidity meters or hygrometers.

9.3.1 Wet and dry bulb hygrometer

When water changes from the liquid state to vapour, i.e. as it evaporates, it requires heat from an external source. The rate of evaporation is dependent on the amount of moisture in the atmosphere. Using these basic facts it is possible to construct a humidity-measuring instrument known as the *wet and dry bulb hygrometer*. It comprises two thermometers, and a simple design is shown in Figure 9.12. Around the bulb of one thermometer is a tight-fitting cotton wick, which dips into a vessel of water. This is termed the *wet bulb thermometer*. The

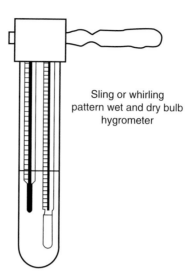

Sling or whirling
pattern wet and dry bulb
hygrometer

Figure 9.12 Sling wet and dry bulb hygrometer.

second thermometer, in an adjacent position, is normal and is known as the *dry bulb thermometer*. The action of the water evaporating from the wick of the wet bulb thermometer cools the bulb, and its temperature, T_w, falls below that of the dry bulb T_D, by an amount depending upon the relative humidity of the air, which is directly related to the partial pressure of the water vapour present in the air and the saturation pressure of water vapour at an atmospheric pressure P_a:

$$P_s = P_{ss} - k \times P_a (T_D - T_w) \qquad [9.6]$$

Rearranging the above equation gives:

$$\frac{P_s}{P_{ss}} = 1 - \frac{k \times P_a}{P_s}(T_D - T_w) \qquad [9.7]$$

This ratio × 100 represents the relative humidity, Φ.

Tables based on the above formulae are issued by various manufacturers and the Meteorological Office. These are normally in terms of the wet bulb depression $(T_D - T_w)$ and the dry bulb temperature T_D, i.e. the temperature of the air at which we wish to know the humidity.

9.3.2 Liquid-in-steel hygrometers

For process applications, liquid-in-steel thermometers have been used widely. The basic features of a design utilising this type of thermometer are indicated in Figure 9.13. The two thermometers are situated in ducts B, through which air is drawn by fan C operated by motor D. The wet bulb is covered with a sleeve of material, which dips down into a trough of water below the ducts. The water in the trough must, of course, be maintained at a sufficient level to keep the sleeve saturated. A two-pen recording instrument is shown; one for the dry

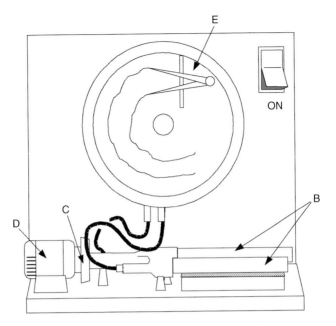

Figure 9.13 Mercury-in-steel wet and dry bulb hygrometer.

bulb and the other for the wet bulb. Relative humidity is then determined from tables based on the depression between the wet and dry bulb temperatures, as described earlier.

9.3.3 Electrical resistance hygrometer

A direct reading electrical resistance wet and dry bulb hygrometer is indicated in Figure 9.14. It comprises two Wheatstone bridges. In the first bridge, there are two thermometers; a wet bulb A and a dry bulb B in opposite branches. The out-of-balance current is a measure of the difference in their resistances and hence in their temperatures. A second dry-bulb thermometer, C, is connected in one arm of the second Wheatstone bridge, and the out-of-balance current here is a measure of the dry-bulb temperature fed to coils of a cross-coil galvanometer. The instrument scale may be calibrated directly in percentage relative humidity based on the electrical signal generated between the two bridges.

9.3.4 Hair hygrometer

Human hair possesses the property of changing its dimensions, in particular the length (dx), with variations in the relative humidity (Φ) in the air and the air temperature (T, in °C), as follows:

$$dx = k.T.\log 10 \frac{100}{\phi} \qquad [9.8]$$

where k is a constant.

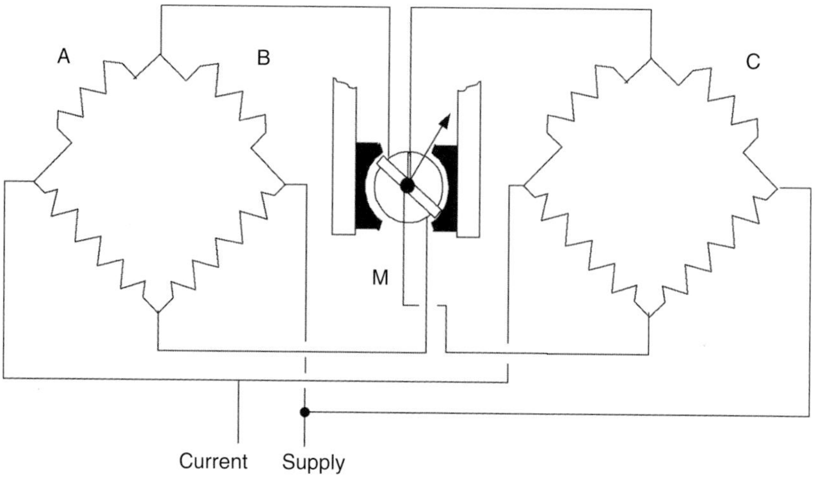

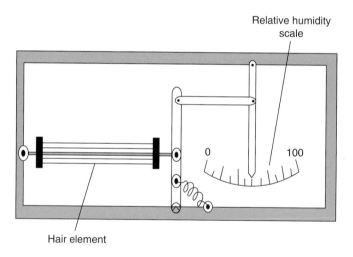

Figure 9.14   Electrical resistance wet and dry bulb direct-reading hygrometer.

Relative humidity
scale

Hair element

Figure 9.15   Hair hygrometer.

The logarithmic relation is unfortunate in the sense that the instrument scale tends to become rather contracted at higher humidity values. Figure 9.15 shows a simple hair hygrometer movement.

9.3.5   Thermal conductivity hygrometer

Heat loss from a heated wire surrounded by gas in a chamber or cell can take place by conduction, convection and/or radiation. If one could arrange by design for the loss to be wholly by conduction, then the rate of heat loss would vary solely with the thermal conductivity of the gas. The conductivity is greatly influenced by the concentration of moisture in the gas, and this suggests a method of measuring absolute humidity.

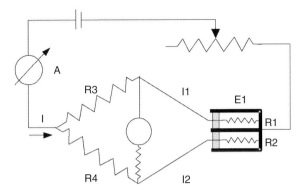

Figure 9.16   Thermal conductivity hygrometer.

Consider Figure 9.16. R3 and R4 are resistances forming two arms of a Wheatstone bridge. R1 and R2 are loops or spirals of a wire of identical resistance, such as platinum, contained in similar cells. The first cell contains the gas, the humidity of which is to be measured (the measuring cell) and the other cell contains the same gas in a saturated or dry condition (the reference cell). A difference in moisture content of the gases in the two cells leads to a difference in temperatures and in resistances of the loops or spirals. The bridge is balanced at a datum value of moisture content and the out-of-balance current is, therefore, a measure of the moisture content at any other value.

9.3.6   Capacitive humidity sensors

This type of meter (Figure 9.17) consists of a substrate on which a thin film of polymer or metal oxide is deposited between two conductive electrodes. The sensing surface is coated with a porous metal electrode to protect it from contamination and exposure to condensation. The substrate is typically glass,

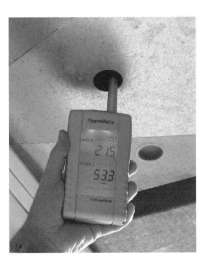

Figure 9.17   Capacitive humidity meter.

ceramic or silicon. The incremental change in the dielectric constant of a capacitive humidity sensor is nearly directly proportional to the relative humidity of the surrounding environment. The change in capacitance is typically 0.2–0.5 pF for a 1% RH change, while the bulk capacitance is between 100 and 500pF at 50% RH at 25°C. Capacitive sensors are characterised by low temperature coefficients, an ability to function at high temperatures (up to 200°C), full recovery from condensation and reasonable resistance to chemical vapours. The typical uncertainty of capacitive sensors is ±2% RH from 5% to 95% RH with two-point calibration.

9.4 Pressure measurement

Measurement of the pressure is essential in general engineering, and in particular in fluid flow applications associated with buildings' energy systems, water supply pumps and so on. In this section, three types of instrument will be discussed, namely barometers, pressure gauges and manometers.

9.4.1 Barometers

Barometers are instruments used to indicate the ambient pressure of air, which is the result of the local pressure exerted by the column of air extending from the ground and up to the limit of the envelope of air in that particular locality. There are two types of barometer: mercury barometers and aneroid barometers.

Mercury barometers

The basic barometer consists of a vertical glass tube which is closed at the top, filled with mercury and stands in a mercury bath. There is a space at the top of the tube in which a vacuum exists, and the height of the column is a measure of atmospheric pressure. The so-called 'Fortin barometer' is a mercury barometer with a Vernier scale (Figure 9.18).

Aneroid barometers

Here, a sealed flexible metal bellows or capsule with a very low internal pressure is connected to a lever with pointer and scale. Atmospheric pressure variations cause a corresponding deflection of the capsule and movement of the pointer. The pointer usually carries a pen, which records the temperature on a rotating chart (Figure 9.19).

9.4.2 Bourdon pressure gauge

In the Bourdon gauge, a curved flattened metal tube is closed at one end and connected to the pressure source at the other end (Figure 9.20). Under pressure the tube tends to straighten and causes a deflection of a pointer through a lever and rack and pinion amplifying system. This gauge can be

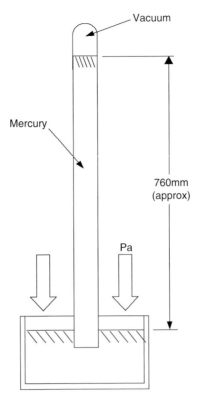

Vacuum

Mercury

760mm
(approx)

Pa

Figure 9.18   Fortin barometer.

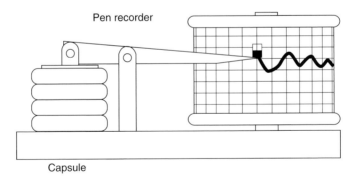

Pen recorder

Capsule

Figure 9.19   Aneroid barometer.

used for liquids or gases from a fraction of a bar pressure up to 10 000 bar. Calibration is by means of a 'dead-weight tester'.

9.4.3 Pressure transducers

A wide range of transducers is available to convert the deflection of a diaphragm or Bourdon tube into an electrical signal which gives a reading on an indicator or is used to control a process, etc. Transducers cover a wide range

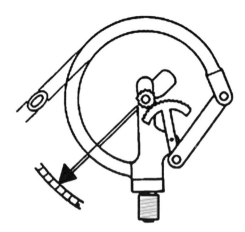

Figure 9.20 Bourdon gauge – mechanism.

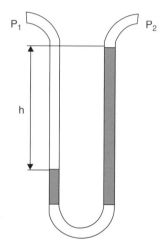

Figure 9.21 The U-tube manometer.

of pressure and have a fast response. Types include piezo-crystal, strain gauge, variable capacity and variable inductance.

9.4.4 Manometers

The U-tube manometer (Figure 9.21) may be used to measure a pressure relative to atmospheric pressure, or the difference between two pressures. The pressure difference is:

$$P_2 - P_1 = (\rho_m - \rho_f)\, g\, h \qquad\qquad [9.9]$$

where:

P is the fluid pressure at points 1 and 2.
ρ is the density of the fluids, flowing (f) and manometer (m) in units of (kg/m³).

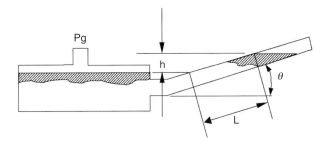

Figure 9.22   The inclined manometer.

g is the gravitational acceleration ($9.81\,m/s^2$)
h is the height of the manometer column in metres.

The inclined single-leg manometer (Figure 9.22) gives greater accuracy, as $\sin(\theta)$ is less than 1. It makes use of a greater length of deflection, and hence a lower likelihood of observer error in reading the pressure.

$$h = L \times \sin(\theta)$$ [9.10]

9.5   Flow measurement

Flow measurements are important in determining the energy flow, as energy is directly proportional to mass flow rate. In this section the different types of flow measurement are described.

9.5.1   Flow measurement by collection

The volume flow rate is measured by recording the time for a given volume of liquid to be collected in a bucket or a measuring tank, see Figures 9.23 and 9.24. The volume flow rate is the ratio of volume over time.

9.5.2   Flow measurement by rotameter

The rotameter (Figure 9.25) is a type of variable-orifice meter consisting of a vertical glass tapered tube containing a metal 'float'. The fluid, which may be a liquid or gas, flows through the annular space between the float and the tube. As the flow is increased, the float moves to a greater height. The movement is roughly proportional to flow, and calibration is usually carried out by the supplier. Angled grooves in the rim of the float cause rotation and give the float stability.

9.5.3   Flow measurement by turbine flow meter

An axial or tangential impeller mounted in a pipe rotates at a speed roughly proportional to the velocity, and hence the flow, of the fluid in the pipe (Figure 9.26). The rotational speed is measured either mechanically or electronically to give flow or flow rate.

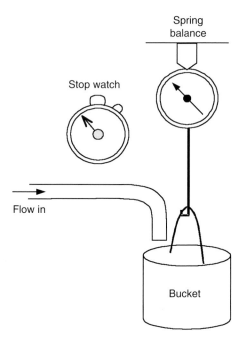

Figure 9.23 Bucket and watch flow measurement.

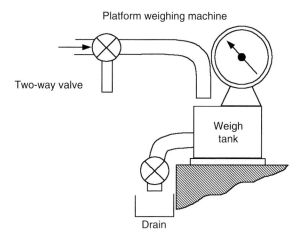

Figure 9.24 Weigh tank flow measurement.

9.5.4 Flow measurement by differential pressure flow meter

There are various methods of velocity/flow rate measurement utilising the pressure drop when the fluid encounters a change in pressure either by an obstruction, such as an orifice, or by bringing part of the flow to rest, such as in the Pitot static tube.

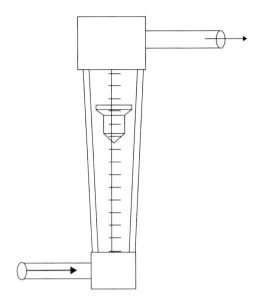

Figure 9.25   Rotameter.

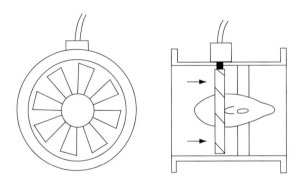

Figure 9.26   Axial turbine meter.

The Pitot tube

The Pitot tube (named after Henri Pitot in 1732) measures fluid velocity by converting the kinetic energy of the flow into potential energy. The conversion takes place at the stagnation point, located at the Pitot tube entrance (see the schematic in Figure 9.27). A pressure higher than the free-stream pressure (dynamic – hole parallel to the flow direction) results from the kinetic to potential energy conversion. This 'static' pressure (hole which is normal to the flow direction) is measured by comparing it to the flow's dynamic pressure with a differential manometer.

Consider two different points along a stream line. The Bernoulli equation yields:

$$\frac{v_1^2}{2g} + z_1 + \frac{p_1}{\rho g} = \frac{v_2^2}{2g} + z_2 + \frac{p_2}{\rho g}$$

[9.11]

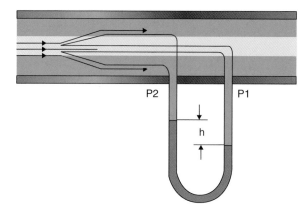

Figure 9.27   Pitot static tube.

If the Pitot static tube is mounted horizontally, then $z_1 = z_2$. Since point 2 is static, i.e. it is a stagnation point, $v_2 = 0$. The approach velocity of the flow can be calculated as:

$$v_1 = \sqrt{\frac{2(p_2 - p_1)}{\rho}}$$

[9.12]

Orifice meter

An orifice meter refers to a thin plate with a central small hole which is placed in the middle of a duct for which the flow rate needs to be measured. When the fluid reaches the orifice plate, it is forced to converge to go through the small hole; the point of maximum convergence actually occurs shortly downstream of the physical orifice, at the so-called *vena contracta point* (see Figure 9.28). As it does so, the velocity and the pressure change. Beyond the vena contracta point, the fluid expands and the velocity and pressure change once again. By measuring the difference in fluid pressure between the normal pipe section and at the vena contracta point, the velocity of flow can be obtained from Bernoulli's equation.

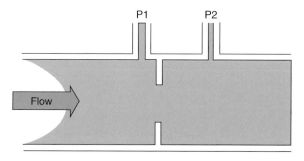

Figure 9.28   Orifice plate.

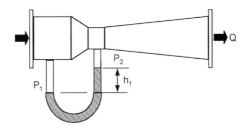

Figure 9.29   Venturi meter.

The conservation of energy between the orifice plate (1) and the full-size duct (2) is given as:

$$P_1 - P_2 = \frac{1}{2} \cdot \rho_1 \cdot V_2^2 - \frac{1}{2} \cdot \rho_1 \cdot V_1^2$$

where the duct is horizontal, the potential difference between sections 1 and 2 is zero.

Introducing the beta factor, $\beta = d_2/d_1$, as well as the coefficient of discharge C_d, which allows for friction and interruption of the flow by the presence of the orifice plate in the flow, we get:

$$V_1 = C_d \times \sqrt{\frac{1}{1-\beta^4}} \sqrt{2(P_1 - P_2)/\rho_1} \qquad \text{[9.13]}$$

Note that C_d for the orifice = 0.6 if manufactured according to British Standards.

Venturi meter

The Venturi meter works in a similar manner to the orifice meter, with the exception that the change in pressure in the Venturi meter is gentler than the orifice plate, as the Venturi meter contains a gradually converging duct (shown in Figure 9.29). Therefore, the same set of equations can be used, with one difference – the value of the coefficient of discharge, C_d, for the Venturi meter has a value in the region of 0.95 and above.

9.5.5   Velocity and flow measured by anemometers

Various types of anemometer are used to measure the velocity of air:

- The *cup-type* (Figure 9.30) is used for free air and has hemispherical cups on arms attached to a rotating shaft. The shape of the cups gives a greater drag on one side than the other and results in a speed of rotation approximately proportional to the air speed. Velocity is found by measuring revolutions over a fixed time.
- The *vane anemometer* (Figure 9.31) has an axial impeller attached to a handle with extensions and an electrical pick-up which measures the revolutions. A meter with several ranges indicates the velocity.

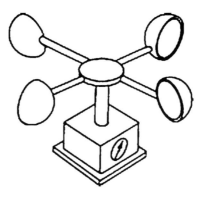

Figure 9.30 Cup-type anemometer.

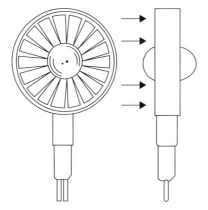

Figure 9.31 Vane anemometer.

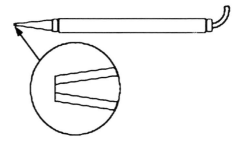

Figure 9.32 Hot-wire anemometer.

● The *hot-wire anemometer* (Figure 9.32) is a probe terminating in an extremely small heated wire element. When subjected to a fluid stream, it cools to an extent depending on the velocity of the fluid passing. The resulting change in resistance of the element is measured by a bridge circuit and is related to velocity by calibration.

Table 9.3 Electrical measurements.

Quantity	Symbol	Unit of measurement	Unit abbreviation
Current	I	Amp	A
Voltage	V	Volt	V
Resistance	R	Ohm	Ω
Power/Energy	E	Watt	W

9.6 Electrical measurements

9.6.1 Energy in electrical circuits

In order to control electrical energy consumption for motors, machines, lights, etc. it is necessary to be able to measure it either directly, as energy in kWh, or indirectly by measuring the voltage, current and phase angle for A/C circuits. Some common electrical measurement information is given in Table 9.3.

As electrical power is the product of voltage and current, the power dissipated in a circuit is the same whether the circuit contains high voltage and low current or low voltage and high current flow. Generally, power is dissipated in the form of heat (heaters), mechanical work (motors, etc.) or energy in the form of radiated (lamps) or stored (batteries) energy.

Electrical energy is measured in Watts together with the time in seconds, with the unit of energy given as Watt-seconds or Joules.

Although electrical energy is measured in Joules, the value can become very large when used to calculate the energy consumed by a component. For example, a single 100W light bulb connected for one hour will consume a total of 100 Watts $\times$ 3600 seconds = 360 000 Joules. So prefixes such as kilo (kJ = 10^3 J) or mega (MJ = 10^6 J) are used to keep the numbers more manageable. If the electrical power is measured in kilowatts and the time is given in hours, then the unit of energy is in kilowatt-hours or kWh, which is commonly called a *unit of electricity* and is what consumers purchase from their electricity suppliers.

9.6.2 Ohm's law

Ohm's law states that the current (I) through a conductor between two points is proportional to the potential difference (V) and inversely proportional to the resistance (R) (see Figure 9.33).

$$V = I \times R \qquad\qquad [9.14]$$

9.6.3 Electrical power

Electrical power (E) in a circuit is the amount of energy that is absorbed or produced within the circuit. The unit of measurement of power is the Watt (W), with prefixes used to denote fractions or large multiples of Watts – e.g. milliwatts (mW = 10^{-3} W) and kilowatts (kW = 10^3 W).

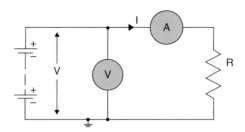

Figure 9.33 Ohm's law.

By using Ohm's law and substituting for V (volts), I (amps) and R (Ω), the formula for electrical power, E (Watts) can be found as:

$$E = V \times I \qquad\qquad [9.15a]$$

$$E = V^2 \div R \qquad\qquad [9.15b]$$

$$E = I^2 \times R \qquad\qquad [9.15c]$$

9.6.4 Alternating current power

The instantaneous electric power in an AC circuit (and also in a DC circuit) is given by:

$$E = VI$$

But these quantities are varying continuously. Almost always the desired power in an AC circuit is the average power, which is given by:

$$E_{avg} = VI \cos\varphi \qquad\qquad [9.16]$$

where φ is the phase angle between the current and the voltage, and where V and I are understood to be the effective, or rms (see below), values of the voltage and current, see Figure 9.34. The term $\cos\varphi$ is called the *power factor* for the circuit; a power factor of one or 'unity power factor' is the goal of any electric utility company, since if the power factor is less than one, it has to supply more current to the user for a given amount of power use. In so doing, the company incurs more line losses. It also must have larger capacity equipment in place than would be necessary otherwise.

The difference between the maximum and minimum values is called peak-to-peak voltage (V_{pp}) and is twice the peak voltage (V_p). The rms voltage (V_{rms}) is related to the peak voltage by:

$$V_{rms} = 0.707 \times V_p \qquad\qquad [9.17]$$

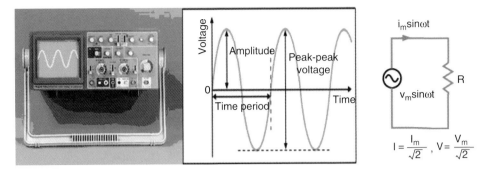

Figure 9.34 Resistive AC circuit.

Circuit currents and voltages in AC circuits are generally stated as root-mean-square, or rms, values rather than by quoting the maximum values. The root-mean-square for a current is defined by:

$$I_{rms} = \sqrt{I^2_{avg}}$$ [9.18]

That is, you take the square of the current and average it, then take the square root. When this process is carried out for a sinusoidal current:

$$[I_m^2 \times \sin^2(\omega.t)]_{avg} = \frac{I_m^2}{2}$$

hence [9.19]

$$I_{rms} = \sqrt{I^2_{avg}} = \frac{I_m}{\sqrt{2}}$$

Since the AC voltage is also sinusoidal, the form of the rms voltage is the same. These rms values are just the effective values needed in the expression for average power:

$$E_{avg} = \frac{V_m \times I_m}{2} \times \cos(\varphi) = V_{rms} \times I_{rms} \times \cos(\varphi)$$ [9.20]

Since the voltage and current are both sinusoidal, the power expression can be expressed in terms of the squares of sine or cosine functions, and the average of a sine or cosine squared over a whole period is 1/2.

9.6.5 Electrical measurements

Energy-related measurements concerned with electricity consumption are usually derived from three elements, namely the electrical current, voltage and resistance of the load. These three variables are directly linked with the energy

Figure 9.35 Basic electrical measurement.

or power consumption of the load and hence it is common to see a single meter combining these to give a direct reading of the energy consumption.

Multimeters are very useful test instruments. By operating a multi-position switch on the meter, they can be quickly and easily set to behave as a *volt-meter*, an *ammeter* or an *ohmmeter*. They have several settings (called *ranges*) for each type of meter and the choice of AC or DC. Multimeters (Figure 9.35) must have a high sensitivity of at least 20 kΩ/V otherwise their resistance on DC voltage ranges may be too low to avoid upsetting the circuit under test and consequently giving an incorrect reading.

As far as power is concerned, the most common unit of power consumption measurement on the electricity meter is the kilowatt-hour, which is equal to the amount of energy used by a load of one kilowatt over a period of one hour, or 3 600 000 joules. Some electricity companies use the SI mega joule instead. Modern electricity meters operate by continuously measuring the instantaneous voltage (Volts) and current (amperes) and finding the product of these to give instantaneous electrical power (Watts) which is then integrated against time to give energy used (joules, kilowatt-hours, etc.). The meters fall into two basic categories – electromechanical and electronic – as shown in Figure 9.36.

The mechanical electricity meter has every other dial rotating counter-clockwise.

Figure 9.36   Electrical power measurement.

The most common type of electricity meter is the Thomson or electromechanical induction watt-hour meter, invented by Elihu Thomson in 1888. This
works by counting the revolutions of an aluminium disc which is made to
rotate at a speed proportional to the power. The metallic disc is acted upon by
two coils. One coil is connected in such a way that it produces a magnetic flux
in proportion to the voltage and the other produces a magnetic flux in proportion to the current. The field of the voltage coil is delayed by 90 degrees using
a lag coil. This produces eddy currents in the disc and the effect is such that a
force is exerted on the disc in proportion to the product of the instantaneous
current and voltage.

A modern digital electronic wattmeter/energy meter samples the voltage
and current thousands of times a second. The average of the instantaneous
voltage multiplied by the current is the true power. The true power divided by
the apparent volt-amperes (VA) is the power factor. A computer circuit uses
the sampled values to calculate the rms voltage, rms current, VA, power
(Watts), power factor and kilowatt-hours. The simpler models display that
information on an LCD. More sophisticated models retain the information over
an extended period of time and can transmit it to field equipment or a central
location.

9.7 Worked examples

Worked example 9.1

In order to study the relationship between the current (*I*), voltage (*V*) and resistance (*R*) as related by Ohm's law, see the circuit shown in Worked example 9.1, Figure 1. Calculate:

(a) The current (*I*) in the circuit for given values of voltage ($V = 12\,V$) and resistance ($R = 3$ ohms).
(b) The resistance (*R*) in the circuit, given values of voltage ($V = 24\,V$) and current ($I = 4$ Amps).
(c) The voltage supplied by a battery, given values of current ($I = 2$ Amps) and resistance ($R = 6$ ohms). Determine the power consumption of the lamp in this case.

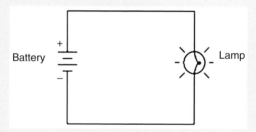

Worked example 9.1, Figure 1

Solution:

(a) Using Ohm's law, the current:

$$I = \frac{V}{R} = \frac{12}{3} = 4\text{ Amps}$$

(b) The resistance in the circuit is:

$$R = \frac{V}{I} = \frac{24}{4} = 6\Omega$$

(c) The voltage provided by the battery is:

$$V = I \times R = 2 \times 6 = 12\text{ Volts}$$

The power consumption of the lamp is:

$$E = I^2.R = 2^2 \times 6 = 24\text{ Watts}$$

Worked example 9.2

Determine the currents associated with each branch of the circuit shown in Worked example 9.2, Figure 1, and the power of the battery.

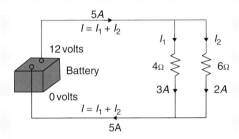

Worked example 9.2, Figure 1

Solution:

Remember that, in parallel branching, the voltage in each branch is the same as that in the main battery.

$$I_1 = \frac{V}{R} = \frac{12}{4} = 3 \text{ Amps}$$

$$I_2 = \frac{V}{R} = \frac{12}{6} = 2 \text{ Amps}$$

Remember that, in parallel branching, the current in each branch depends on the resistance of the branch; the total current in the battery is the sum of both currents flowing in each branch.

Hence, the total current flowing through the battery

$$I = 3 + 2 = 5 \text{ Amps}.$$

The power of the battery is:

$$E = V.I = 12 \times 5 = 60 \text{ Watts}$$

Worked example 9.3

The inclined manometer shown in Worked example 9.3, Figure 1 is initially zeroed when no pressure is applied. If point V is now connected to a known pressure of 100 Pa, while point P is open to the atmosphere, what will the deflection on the manometer scale be if it is inclined at 30° to the horizontal and the manometer density is 850 kg/m³?

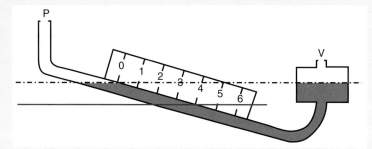

Worked example 9.3, Figure 1

Solution:

The hydrostatic equation when the manometer is in equilibrium implies that the left-hand side and the right-hand side pressures are equal, hence:

$$P = \rho g h$$

Therefore:

$$h = P/\rho.g = 100/(850 \times 9.81) = 0.012 \, \text{m}$$

Applying trigonometry, $h = L \times \sin(30)$. Hence:

$$L = h/\sin(30)$$

$$= 0.012/0.5$$

$$= 0.024 \, \text{m or } 2.4 \, \text{cm}.$$

Worked example 9.4

A Venturi meter fitted in a 15 cm pipeline has a throat diameter of 7.5 cm. The pipe carries water, and a U-tube manometer mounted across the Venturi meter has a reading of 95.2 mm of mercury. Using Worked example 9.4, Figure 1 for reference, determine:

(a) The pressure drop in Pascals indicated by the manometer.
(b) The ideal throat velocity (m/s).
(c) The actual flow rate (L/s) if the meter C_D is 0.975.

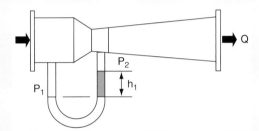

Worked example 9.4, Figure 1

Solution:

$D_1 = 0.15\,\text{m},$ $A_1 = 0.0176\,\text{m}^2$
$D_2 = 0.075\,\text{m},$ $A_2 = 0.00441\,\text{m}^2$

(a)

$$p_1 - p_2 \text{ (Pascals)} = \rho_{manometer} \times g \times h$$
$$= 13\,600 \times 9.81 \times 0.0952 = 12\,701\,\text{Pa}$$

(b)

$$\bar{V} = \sqrt{\frac{2(p_1 - p_2)}{\rho[1 - (A_2 / A_1)^2]}}$$

$$= \sqrt{\frac{2(12701)}{1000[1 - (0.0629)]}} = 5.206\ \text{m/s}$$

(c)

$$\dot{V} = C_d\,\bar{V}_2\,A_2$$
$$= 0.975 \times 5.206 \times 0.00441 = 0.0224\,\text{m}^3/\text{s} = 22.4\,\text{L/s}.$$

9.8 Tutorial problems

9.1 For the circuit shown in Tutorial problem 9.1, Figure 1, find the voltage, V, and the power, P, consumed by the load $R = 12$ ohms.
Ans. (24 V, 48 W)

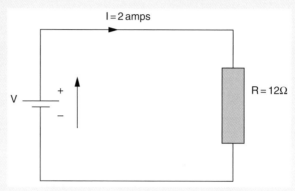

Tutorial problem 9.1, Figure 1

9.2 A string of Christmas tree lights consisting of twenty lamps is connected in series across the 240 V mains supply. The power consumption of the whole string is 24 W. Calculate the resistance of each lamp.
Ans. (120 ohms)

9.3 A section of a river has a horizontal bed and parallel sides. At a point upstream of a narrow section, the depth is 1.6 m and width 8 m. In passing through the narrow section, which is 6.5 m wide, the free surface falls by 0.15 m. Determine:

(a) the flow velocity in each section;
(b) the volumetric flow rate of the river.

Ans. (1.86 m/s, 2.53 m/s, 24 m³/s)

9.4 Kerosene at 20°C flows at 18 m³/h in a 5 cm diameter pipe. If a 2 cm diameter thin plate orifice with corner taps is installed, what will the measured drop be in Pa? Use the following data: $C_d = 0.603$; $\mu = 192\,e^{-3}$ kg/(m s) for kerosene; $\rho = 804$ kg/m³.
Ans. (6.980 kPa)

Chapter 10

Renewable Energy Technology

Learning outcomes

• Demonstrate the various forms of renewable energy	Knowledge and understanding
• Describe solar energy and the relationship between latitude, longitude and the different geometrical variables affecting the incident energy from the sun.	Analysis
• Describe wind energy, the difference between vertical axis, horizontal axis and the various designs available on the market.	Knowledge and understanding
• Describe biomass and its derivatives as renewable fuel	Knowledge and understanding
• Calculate the energy potential from all four renewable sources	Analysis
• Solve problems associated with renewable energy	Problem solving
• Practise further tutorial problems	Problem solving

Energy Audits: A Workbook for Energy Management in Buildings, First Edition.
Tarik Al-Shemmeri.
© 2011 Blackwell Publishing Ltd. Published 2011 by Blackwell Publishing Ltd.

10.1 Introduction

Much attention has been paid recently to 'renewables' as a potential source of fuel. The rising price of oil and the logistics of supplying fossil fuels to remote areas, together with the environmental incentive, are the main factors driving the push towards renewables. In remote locations, stand-alone renewable energy systems can be more cost effective than extending a power line to the electricity grid. In addition, the environmental benefits, given the current international concerns on global warming, make such a project much more valuable and rewarding.

The growth of renewable energy sources also stimulates employment, the creation of new technologies and new skills.

The new EU Directive on renewable energy sets ambitious targets for all Member States, such that the EU will reach a 20% share of energy from renewable sources by 2020 and a 10% share of renewable energy specifically in the transport sector. It also improves the legal framework for promoting renewable electricity, requires national action plans that establish pathways for the development of renewable energy sources including bioenergy, creates cooperation mechanisms to help achieve the targets cost effectively and establishes the sustainability criteria for biofuels.

In a statement just before implementation of the Directive, Ed Miliband, the then UK Secretary of State for Energy and Climate Change, spelled out the Government strategy:

'Transforming the country into a cleaner, greener and more prosperous place to live is at the heart of our economic plans for "building Britain's future" and ensuring the UK is ready to take advantage of the opportunities ahead.'

By 2020:

- More than 1.2 million people will be in green jobs.
- 7 million homes will have benefited from whole house makeovers, and more than 1.5 million households will be supported to produce their own clean energy.
- Around 40% of electricity will be from low-carbon sources, from renewables, nuclear and clean coal.
- We will be importing half the amount of gas that we otherwise would.
- The average new car will emit 40% less carbon than it does now.

Significant environmental benefits can be obtained from using a renewable energy device, attributed with preventing the release of greenhouse gases associated with fossil fuels. The general equation for estimating the reduction in emitted gas is:

Gas-emission reduction (in tonnes) $= A \times 0.8 \times h \times kG$ 　　　　　[10.1] where

> A is the rated capacity of the development in kW
> h is the number of operational hours per year $= 8000h$
> kG is the specific emitted gas constant

Hence, the following equations are used to predict environmental benefits from using 1kWe PV system:

$$CO_2 \text{ emission reduction (in tonnes)} = 1 \times 0.8 \times 8000 \times 862/10^6$$
$$= 5.5$$
$$SO_2 \text{ emission reduction (in tonnes)} = 1 \times 0.8 \times 8000 \times 9.9/10^6$$
$$= 0.063$$
$$NO_2 \text{ emission reduction (in tonnes)} = 1 \times 0.8 \times 8000 \times 2.9/10^6$$
$$= 0.018$$

10.2 Solar energy

The Earth receives more energy from the sun in just one hour than the world's population uses in a whole year. The total solar energy flux intercepted by the Earth on any particular day is 4.2×10^{18} Watthours or 1.5×10^{22} Joules. This is equivalent to burning 360 billion tons of oil per day or 15 billion ton oil equivalent (toe) per hour. In fact, the world's total energy consumption of all forms in the year 2005 was only 10 537 Mtoe (Source: BP Statistical Review of World Energy, 2006).

Solar radiation is a general term for the electromagnetic radiation emitted by the sun. Every location on Earth receives sunlight at least part of the year. The amount of solar radiation that reaches any one 'spot' on the Earth's surface varies according to these factors:

- time of day;
- season;
- geographic location;
- topography;
- local weather.

The Earth revolves around the sun in an elliptical orbit and is closer to the sun during part of the year. When the sun is nearer the Earth, the Earth's surface receives a little more solar energy. The Earth is nearer the sun when it's summer in the southern hemisphere and winter in the northern hemisphere. However, the presence of vast oceans moderates the hotter summers and colder winters one would expect to see in the southern hemisphere as a result of this difference.

The 23.5° tilt in the Earth's axis of rotation is a more significant factor in determining the amount of sunlight striking the Earth at a particular location.

Tilting results in longer days in the northern hemisphere from the spring (vernal) equinox to the autumn (autumnal) equinox and longer days in the southern hemisphere during the other six months. Days and nights are both exactly 12 hours long on the equinoxes, which occur each year on or around March 23 and September 22.

10.2.1 Solar declination

The axis of the Earth's daily rotation around itself is at an angle of 23.45° to the axis of its elliptical orbital plane around the sun. This tilt is the major cause of the seasonal variation in solar radiation available at any location on Earth. The angle between the Earth-sun line and the plane through the Equator is called *solar declination*:

$$\delta = 23.45 \times \sin\left(360 \times \frac{284 + n}{365}\right) \qquad [10.2]$$

10.2.2 Solar altitude angle and azimuth angle

The position of the sun can be described by an altitude angle and an azimuth angle. The solar azimuth angle is the angle between a line collinear with the sun's rays and the horizontal plane. The solar azimuth angle is the angle between a due south line and the projection of the site-to-sun line on the horizontal plane. The solar zenith angle is the angle between the site-to-sun line and the vertical at the site: $z_p = 90° - \alpha$.

10.2.3 Solar time and angles

The sun angles are obtained from the local solar time, which differs from the local standard time. The relationship between local solar time and local standard time (LST) is:

$$\text{Solar time} = \text{LST} + \text{ET} + (\Phi_{st} - \Phi_{local}) \times 4\,\text{min/degree} \qquad [10.3]$$

ET is the equation of time, which is a correction factor that accounts for the irregularity of the speed of the Earth's motion around the sun.

$$E = 9.87 \times \sin(2B) - 7.53 \times \cos(B) - 1.5 \times \sin(B) \qquad [10.4]$$

where

$$B = \frac{360 \times (n - 81)}{364}$$

The *solar altitude* is the angle a direct ray from the sun makes with the horizontal at a particular place on the surface of the Earth (Figure 10.1). The altitude angle can be found using the following formula:

$$\sin \alpha = \sin \Phi \sin \delta + \cos \Phi \cos \delta \cos h_s \qquad [10.5]$$

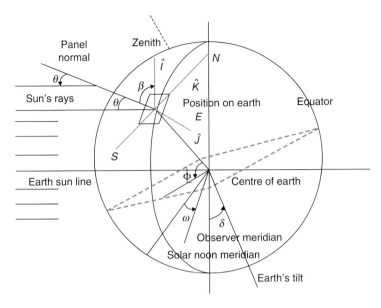

Figure 10.1 Solar definitions.

The *solar azimuth* is the angle the horizontal component of the direct ray from the sun makes with the true south in the northern hemisphere. It is defined as:

$$\sin a_s = \cos \delta \sin \omega_s / \cos \alpha \qquad\qquad [10.6]$$

Sunrise and sunset times (based on the centre of the sun being at the horizon) are defined as:

$$\omega_{ss} \text{ or } \omega_{sr} = \pm \cos^{-1}[-\tan \Phi \tan \delta] \qquad\qquad [10.7]$$

10.2.4 Solar radiation

The Earth revolves around the sun in an elliptical orbit for 365.25 days. The Earth's orbit reaches a maximum distance from the sun, or aphelion, on July 21. The minimum Earth–sun distance, the perihelion, occurs on January 21. The Earth rotates about its own polar axis, which is inclined to the elliptic plane by 23.45°.

This tilt is a major cause of the seasonal variation in solar radiation available at any location on the Earth. The variation in seasonal solar radiation availability on the Earth can be understood from the geometry of the relative movement of the Earth around the sun. Since the Earth's orbit is elliptical, the Earth–sun distance varies during a year; the variation being ±1.7% from the average. Therefore, extraterrestrial radiation, I, also varies by the inverse square law:

$$I = I_0 \, (d_0/d)^2 \qquad\qquad [10.8]$$

where d is the distance between the sun and the Earth and d_0 is the yearly mean Earth-sun distance (1.496×10^{11}m). The $(d_0/d)^2$ factor may be approximated as:

$$(d_0/d)^2 = 1.00011 + 0.034221 \cos X + 0.00128 \sin X + 0.000719 \cos 2X + 0.000077 \sin 2X$$

where $X = 360 \, (n - 1)/365$ and n is the number of days from January 1.

The following approximate relationship may also be used without much loss of accuracy:

$$I = I_0 \, [1 + 0.034 \cos (360n/365.25)] \qquad\qquad [10.9]$$

10.2.5 Incidence angle

The angle between the sun's ray and a vector normal to the aperture or surface of the collector is called the *angle of incidence*. The other angle of importance discussed in this section is the *tracking angle*. Most types of mid- and high-temperature collector require a tracking drive system to align at least one and often both axes of the collector aperture perpendicular to the sun's central ray.

10.2.6 Fixed aperture

An arbitrarily oriented surface or aperture that does not track may be described in terms of the orientation of the collector and solar altitude and azimuth angles. For example, a fixed collector aperture is defined by the tilt angle (Figure 10.2) and aperture azimuth angle:

$$\cos(\theta) = \cos(\alpha_s) \times \cos(\beta) - \sin(\alpha_s) \times \sin(\beta) \times \cos(\gamma_s - \gamma) \qquad [10.10]$$

where

α = altitude angle
β = collector tilt angle
γ_s = collector azimuth variable angle
γ = solar azimuth angle

A fixed horizontal surface receives more aperture irradiation over the day (28% more in Albuquerque) than does a tilted-latitude, south-facing surface in summer. In the winter, however, the horizontal fixed surface in Albuquerque receives only 49% of the daily energy that a tilted-latitude surface does.

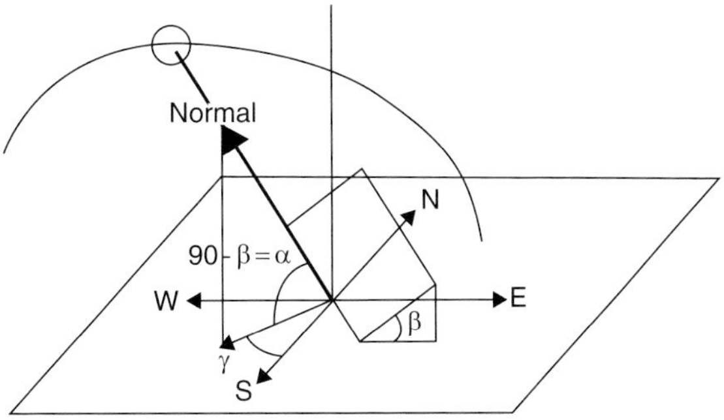

Figure 10.2   Solar angles for a tilted surface.

10.2.7   Solar tracking

The single tracking axis is often horizontal and can be oriented in any direction. If an aperture tracks the sun by rotation about one axis, the beam aperture irradiance is reduced by the cosine of the angle between the tracking axis and the sun:

$$\text{Cos } \theta = 1 \times \cos^2 \alpha \times \cos^2 (\alpha - \beta) \tag{10.11}$$

The maximum amount of insolation is collected when the aperture points directly toward the sun, and hence the angle of incidence is zero. If a collector aperture can be tracked about two independent axes, the angle of incidence can be maintained at zero throughout the day for any day of the year:

$$\text{Cos } \theta = 1$$

10.2.8   The aperture intensity

The aperture intensity is summed over a period of time, usually a full day. Often it is of interest to sum the instantaneous rate of energy falling on a surface over a full time period in order to describe the total amount of energy deposited on the surface from sunrise to sunset. This sum is called the *solar irradiation* (W/m²). Daily irradiation on a horizontal surface may be calculated from the instantaneous rate of energy falling on that surface. Finally, the daily aperture intensity is found by summing the aperture intensity from sunrise to sunset:

$$H = \sum_{i=1}^{24} AI(t) \tag{10.12}$$

The collector aperture irradiance, the rate at which solar energy is incident on the aperture per unit aperture area, may be found from the irradiance parameters.

Table 10.1 Monthly and yearly average insolation in the UK.

Location	Jan	Feb	Mar	Apr	May	June	July	Aug	Sept	Oct	Nov	Dec	Average
Edinburgh Latitude 55' 55" N Longitude 3'0" W	0.44	0.94	1.86	3.18	4.33	4.34	4.13	3.41	2.43	1.2	0.59	0.32	2.26
London Latitude 51' 32" N Longitude 0' 5" W	0.67	1.26	2.22	3.48	4.54	4.51	4.74	4.01	2.86	1.65	0.89	0.52	2.61

Monthly horizontal insolation (kWh/m²/day).
Source: NASA.

The beam component of the insolation is of interest. So the aperture irradiance for beam insolation may be calculated as:

$$AI\ (t) = I \times Cos\ \theta \qquad [10.13]$$

10.2.9   Energy conversion efficiency

A solar cell's energy conversion efficiency (η, 'eta') is the percentage of power converted (from absorbed light to electrical energy) and collected when a solar cell is connected to an electrical circuit. This term is calculated using the ratio of P_m divided by the input light irradiance under 'standard' test conditions (E, in W/m^2) and the surface area of the solar cell (A_c in m^2):

$$\eta = \frac{P_m}{E \times A_c} \qquad [10.14]$$

10.2.10   Installation of photovoltaic modules

The proper tilt and azimuth angle choice is far more important for photovoltaic systems design than solar thermal system design. Manual or automatic tilt angle adjustment can increase the total light–electricity conversion by up to 30%, and more in locations with high values of solar radiation. The incidence angle should be as close to 90° as possible. Photovoltaic module tilt angle and location choice in general require more care than solar collectors' tilt angle and location choice. Shaded locations, including partially shaded areas, are not suitable for photovoltaic module fixation. Modules should be oriented in a southerly direction. Table 10.1 gives the monthly and yearly average insolation at two locations in the UK while Figure 10.3 gives figures for Stafford, UK.

10.2.11   Technology status

Solar energy can be captured by solar panels. There are two main types of solar panel which use completely different technologies to make use of the energy from the sun: solar water-heating collectors and solar electric panels (or photovoltaic cells).

Solar water-heating collectors

Solar water-heating systems are the most popular form of solar energy used in the UK. The system is connected to the hot water system. Solar water-heating systems can provide over half of a household's hot water requirements over the year. There are two types of solar water-heating collector: flat plate and evacuated tubes.

- *Flat plate collectors*. In their simplest form, solar water-heating panels are made from a sheet of metal painted black which absorbs the sun's energy.

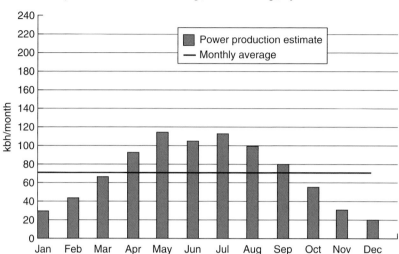

52°48′2"North, 2°6′27"West, nearest city: Stafford, United Kingdom
Non. power = 1 kW, Inclin. = 37 deg., Orient. = 0 deg., System losses = 14.0%

Figure 10.3 Solar energy for Stafford, UK.

Water is fed through the panel in pipes attached to the metal sheet and picks up the heat in the metal. For the UK climate, the pipework contains nontoxic anti-freeze. The pipes are often made of copper for better conduction. The metal sheet is embedded in an insulated box and covered with glass or clear plastic on the front. The system is usually installed on the roof.

● *Evacuated tubes*. The evacuated tube system is a series of glass heat tubes grouped together. The tubes are highly insulated due to a vacuum inside the glass.

The cost of installing a solar hot water system ranges from approximately £500–£1500 for a DIY system to £2000–£5000 for a commercially installed system. These prices, however, are dependent on the size of the system. A typical installation in the UK has a panel of 3 m² to 4 m² with a storage tank of 150–200 L (2 m² for evacuated tubes). However, the optimum size will depend on actual hot water use. This can be calculated using software to simulate system performance throughout the year.

Solar electric panels (photovoltaic cells)

These panels transform the solar radiation directly into electricity. Types of PV technology available on the market today include:

● *Monocrystalline silicon cells*. These cells are made from very pure monocrystalline silicon. The silicon has a single and continuous crystal lattice structure with almost no defects or impurities. The principal advantage of monocrystalline cells is their high efficiency values, typically around 15%, although the manufacturing process required to produce monocrystalline silicon is complicated, resulting in slightly higher costs than other technologies.

- *Multicrystalline silicon cells.* Multicrystalline cells are produced using numerous grains of monocrystalline silicon. In the manufacturing process, molten polycrystalline silicon is cast into ingots; these ingots are then cut into very thin wafers and assembled into complete cells. Multicrystalline cells are cheaper to produce than monocrystalline ones, due to the simpler manufacturing process. However, they tend to be slightly less efficient, with average efficiencies of around 12%.
- *Amorphous silicon cells.* Amorphous silicon cells are composed of silicon atoms in a thin homogenous layer rather than a crystal structure. Amorphous silicon absorbs light more effectively than crystalline silicon, so the cells can be thinner. For this reason, amorphous silicon is also known as a *thin film PV technology*. Amorphous silicon can be deposited on a wide range of substrates, both rigid and flexible, which makes it ideal for curved surfaces and 'fold-away' modules. Amorphous cells are, however, less efficient than crystalline-based cells, with typical efficiencies of around 6%, but they are easier and therefore cheaper to produce.

10.2.12 PV system components

A typical PV system is shown in Figure 10.4; the system has four basic components.

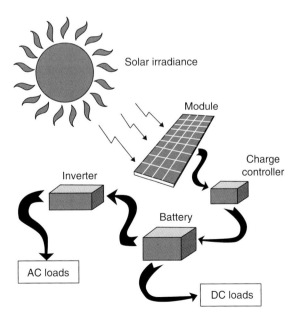

Figure 10.4   Layout of a PV system.

The PV array

This consists of a number of individual photovoltaic modules connected together to give a suitable current and voltage output. Common power modules have a rated power output of around 50–250 Watts each. A typical domestic system of 1.5–2 kWp may therefore comprise many modules covering an area, depending on the technology used and the orientation of the array with respect to the sun.

Most power modules deliver direct current (DC) electricity at 12 Volts, whereas most common household appliances run off alternating current (AC) at either 110 V or 230 V. An inverter is used to convert the low-voltage DC to higher voltage AC. Numerous types of inverter are available, but not all are suitable for use when feeding power back into the main electrical supply grid.

Solar cell operating characteristics

Figure 10.5 shows that with constant irradiance, the output voltage of a cell or an array of cells falls as it is called upon to deliver more current. Maximum power delivery occurs when the voltage has dropped to about 80% of the open-circuit voltage.

The *fill factor (FF)* is defined as the ratio between the power at the maximum power point and the product of the open-circuit voltage and short-circuit current. It is typically better than 75% for good quality solar cells.

Figure 10.6 shows the performance of a 17 Volt, 4.4 Amp, 75 Watt PV array used to top up a 12 Volt battery. If the actual battery voltage is 12 Volts, the resulting current will only be about 2.5 Amps and the power delivered by the array will be just over 50 Watts rather than the specified 75 Watts: an efficiency loss of over 30%. Maximum power point tracking is designed to overcome this problem.

The power tracker module is a form of voltage regulator which is placed between the PV array and the battery. It presents an ideal load to the PV array,

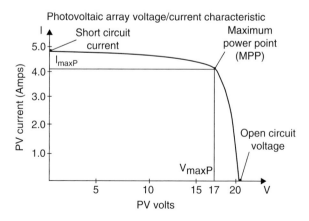

Figure 10.5   PV cell performance curve.

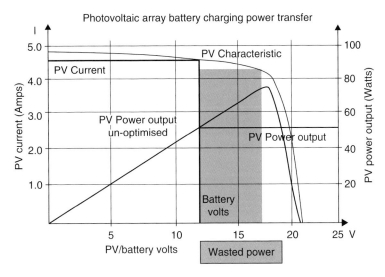

Figure 10.6 PV cell performance curve.

allowing it to operate at its optimum voltage, in this case 17 Volts, delivering its full 75 Watts regardless of the battery voltage. A variable DC/DC converter in the module automatically adjusts the DC output from the module to match the battery voltage of 12 Volts. As the voltage is stepped down in the DC/DC converter, the current will be stepped up in the same ratio. Thus, the charging current will be 17/12 × 4.4 = 6.2 Amps and, assuming no losses in the module, the power delivered to the battery will be 12 × 6.23 = the full 75 Watts generated by the PV array. In practice, the converter losses could be as high as 10%. Nevertheless, a substantial efficiency improvement is possible.

Batteries

Batteries store the energy created by solar panels, so that power can be used when the sun isn't shining. Batteries are generally 80% efficient.

The 'deep-cycle' (generally lead–acid) batteries typically used for small systems last five to ten years and reclaim about 80% of the energy channelled into them. In addition, these batteries are designed to provide electricity over long periods, and can repeatedly charge and discharge up to 80% of their capacity. Automotive batteries, which are shallow-cycle (and therefore prone to damage if they discharge more than 20% of their capacity), should not be used.

The cost of deep-cycle batteries depends on the type, capacity and climate conditions under which they will operate, the frequency of maintenance and chemicals used to store and release electricity. For safety, batteries should be located in a space that is well ventilated and isolated from living areas and electronics, as they contain dangerous chemicals and emit hydrogen and oxygen gas while being charged. In addition, the space should provide protection from temperature extremes. Be sure to locate your batteries in a space that has easy access for maintenance, repair and replacement. Batteries can be recycled when they wear out.

Charge controller

This device regulates rates of flow of electricity from the generation source to the battery and the load. The controller keeps the battery fully charged without overcharging it. When the load is drawing power, the controller allows the charge to flow from the generation source into the battery, the load or both. When the controller senses that the battery is fully (or nearly fully) charged, it reduces or stops the flow of electricity from the generation source, or diverts it to an auxiliary or 'shunt' load (most commonly an electric water heater).

Many controllers will also sense when loads have taken too much energy from batteries and will stop the flow until sufficient charge is restored to the batteries. This last feature can greatly extend the battery's lifetime.

Inverter

An inverter changes the type of power (low-voltage DC to high-voltage AC), so that we can run common appliances. Inverters condition electricity so that it matches the requirements of the load. If you plan to tie your system to the electricity grid, you will need to purchase conditioning equipment that can match the voltage, phase, frequency and sine wave profile of the electricity produced by your system to that flowing through the grid.

Additional equipment

Safety features protect stand-alone small renewable energy systems from being damaged or harming people. Here are the major safety features a PV system will need:

- safety disconnects;
- grounding equipment;
- surge protection;
- meters and instrumentation.

10.3 Wind energy

According to BWEA, the British Wind Energy Authority, in the UK there are currently 2896 large wind turbines with installed capacity of 4532 MW, sufficient to supply over 2.5 million homes (based on annual household energy consumption of 4.7 MWh). The rapid growth of this form of energy since the turn of the millennium is charted in Figure 10.7, and the positive impact on the environment from these can be seen from the emission reductions shown in Table 10.2.

The placement or 'siting' of wind systems is extremely important. In order for a wind-powered system to be effective, a relatively consistent wind flow is

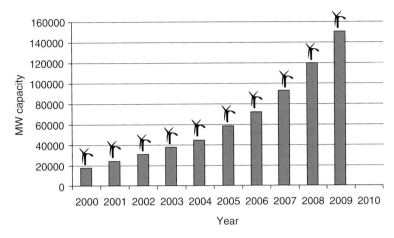

Figure 10.7   Wind power development.

Table 10.2   Wind energy savings on gas emissions (BWEA).

Gas	Reduction Tonnes per year
CO_2	5 121 273
SO_2	119 099
NOx	35 730

Source: http://www.bwea.com/ukwed/index.asp

required. Obstructions such as trees or hills can interfere with the rotors. Because of this, the rotors are usually placed atop towers to take advantage of the stronger winds available higher up. Furthermore, wind speed varies with temperature, season and time of day. All these factors must be considered when choosing a site for a wind-powered generator.

The amount of wind energy available at any location depends on two sets of factors:

● Climatic factors including: time of day, season, geographic location, topo graphy and local weather.
● Mechanical factors including: diameter of rotor and type of turbine.

10.3.1   Ideal wind power calculation

In theory, wind energy is calculated by the following general equation (the proof for which will be derived in the following section):

$$P = C_p \times 1/2 \times \rho \times A \times V^3 \tag{10.15}$$

where

    P is the shaft power (W)

    C_p is the rotor coefficient (ratio of the shaft power of the windmill to the power in the wind in the cross-sectional area of the rotor)

    ρ is the density of the oncoming air

    V is the velocity of the wind (m/s)

The actual power is further reduced by two more inefficiencies, due to the gear box losses and the generator efficiency.

The value of the ideal power is limited by what is known as the Betz coefficient, which has a value of $C_p = 0.59$ as the highest possible conversion efficiency. In practice, most wind turbines have efficiencies well below 0.5, depending on the type, design and operational conditions.

In the operational output range, wind power generated increases with wind speed cubed. In other words, at a wind speed of 5 m/s, the power output is proportional to $5^3 = 125$, whereas at a wind speed of 10 m/s, the power output is proportional to 1000. Doubling the speed from 5 to 10 m/s therefore results in a power increase of eightfold. This highlights the importance of location when it comes to installing wind turbines.

The rotor diameter affects the power output in a square manner, i.e., doubling the rotor diameter results in increasing the power output by four times. On the other hand, since power generated is related to wind speed by a cubic ratio, if your turbine is rated at producing 1 kW at 12 m/s then it will produce 125 W at 6 m/s and 15 W at 3 m/s.

10.3.2 Theory of wind turbines

A windmill extracts power from the wind by slowing down the wind (Figure 10.8). At stand still, the rotor obviously produces no power, while at very high rotational speeds, the air is more or less blocked by the rotor and again no power is produced.

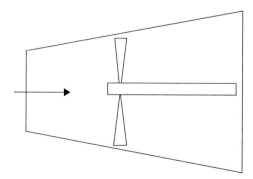

Figure 10.8 Ideal wind energy theory.

The power produced (P_{kin}) by the wind turbine is the net kinetic energy change across the wind turbine (from initial air velocity of V_1 to a turbine exit air velocity of V_2) and is given as:

$$P_{kin} = 1/2 \times m \times [(V_1)^2 - (V_2)^2] \qquad [10.16]$$

The mass flow rate of wind is given by the continuity equation as the product of density, area swept by the turbine rotor and the approach air velocity:

$$m = (\rho \times A \times V_a) \qquad [10.17]$$

hence the power becomes:

$$P_{kin} = 1/2 \times (\rho \times A \times V_a) \times [(V_1)^2 - (V_2)^2] \qquad [10.18]$$

Since the rotor speed is the average speed (V_a) between inlet and outlet:

$$V_a = 1/2 \times (V_1 + V_2) \qquad [10.19]$$

Hence, the power is

$$
\begin{aligned}
P_{kin} &= (1/2) \times \rho \times A \times (V_1 + V_2)/2) \times [(V_1)^2 - (V_2)^2] \\
&= (1/4) \times \rho \times A \times [V_1^3 - V_2^3 - V_1 \times V_2^2 + V_1^2 \times V_2] \\
&= (1/4) \times \rho \times A \times V_1^3 \times [1 - (V_2/V_1)^3 - (V_2/V_1)^2 + (V_2/V_1)] \qquad [10.20]
\end{aligned}
$$

To find the maximum power extracted by the rotor, differentiate Equation [10.20] with respect to V_2 and equate it to zero:

$$dP_{kin}/dV_2 = 1/4 \times \rho \times A \times (-3 \times V_2^2 - 2 \times V_1 \times V_2 + V_1^2) = 0 \qquad [10.21]$$

Since the area of the rotor (A) and the density of the air (ρ) cannot be zero, the expression in the bracket of Equation [10.21] has to be zero. Hence, the quadratic equation becomes:

$$(3V_2 - V_1) \times (V_2 + V_1) = 0$$

Since $V_2 = -V_1$ is unrealistic in this situation, there is only one solution. Equation [10.21] yields:

$$V_2 = 1/3 \times V_1 \qquad [10.22]$$

Substitution of Equation [10.22] into Equation [10.20] results in:

$$P_{kin} = (0.5925) \times 1/2 \times \rho \times A \times V_1^3 \qquad [10.23]$$

The theoretical maximum fraction of the power in the wind which could be extracted by an ideal windmill is, therefore, the fraction 0.5925 and is called

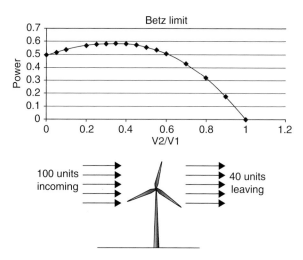

Figure 10.9 Betz limit on wind energy efficiency and its implications.

the *Betz coefficient*. Because of aerodynamic imperfections in any practical machine and mechanical losses, the power extracted is less than that calculated above. Figure 10.9 demonstrates the effect of wind turbine design implications on the resulting power that can be harnessed from the incoming wind. Efficient wind turbines depend on the production of that optimum speed ratio, giving the maximum or near to the maximum power possible.

Equation [10.23] clearly shows that:

- The power is proportional to the density (ρ) of the air, which varies slightly with altitude and temperature.
- The power is proportional to the area (A) swept by the blades and thus to the square of the radius (R) of the rotor.
- The power varies with the cube of the wind speed (V^3). This means that the *power increases eightfold if the wind speed is doubled*. Hence, one has to pay particular attention to site selection.

There are two further factors to be considered when estimating the power output from a turbine: the first is the mechanical transmission and the second is the generator's efficiency, both of which are less than unity, hence the real power is proportionately less than the ideal value calculated in Equation [10.23].

The ratio of actual productivity in a year to the theoretical maximum (total hours in a year) is called the *capacity factor, C_f*. Assuming a 5 kW wind turbine generates 10 MWh annually, if that same installation had run – theoretically – for 24 hours a day and 365 days a year at full load, it would have generated 43.8 MWh. The capacity factor (C_f) is 10/43.8 = 0.23. Typical values for C_f lie between 0.2 and 0.4 in the United Kingdom, depending on the exact location.

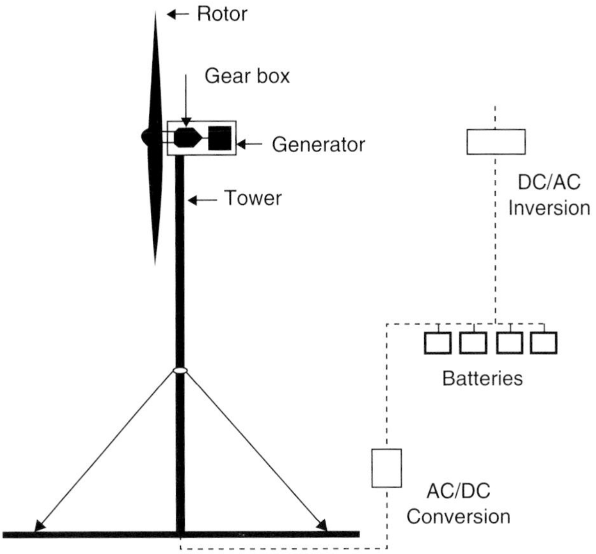

Figure 10.10   Wind energy system.

10.3.3   Wind turbine components

A wind turbine usually has six main components: the rotor, the gearbox, the generator, the control and protection system, the tower and the foundation. These main components can be seen in Figure 10.10.

The *rotor* takes the wind and aerodynamically converts its energy into mechanical energy through a connected shaft.

The *gearbox* increases the rotational velocity of the shaft for the generator. In some turbines, the gearbox is not needed because the rotational velocity or the torque from the shaft is high enough.

The *generator* is a device that produces electricity when mechanical work is given to the system.

The *control and protection system* is like a safety feature that makes sure that the turbine will not work under dangerous conditions.

The *tower* is the main shaft that connects the rotor to the foundation. It also raises the rotor high in the air where we can find stronger winds.

The *foundation,* or base, supports the entire wind turbine and makes sure that it is well fixed to the ground.

In addition to the main components of a wind turbine, a wind energy generation system is shown in Figure 10.10, incorporating a charge controller, battery and inverter.

10.3.4   Types of wind turbine

There are two main types of wind turbine: horizontal axis and vertical axis. The horizontal axis wind turbine (HAWT) and the vertical axis wind turbine (VAWT) are classified or differentiated by the axis of rotation of the rotor shafts.

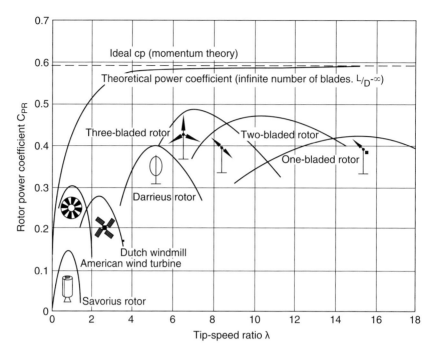

Figure 10.11   Typical wind turbine efficiencies. Courtesy of http://www.scielo.cl/fbpe/img/ingeniare/v17n3/fig06-2.JPGwww.scielo.cl

Horizontal axis wind turbines

Horizontal axis wind turbines, often shortened to HAWT-type turbines, or just HAWTs, have a horizontal rotor shaft and an electrical generator, both of which are located at the top of a tower. This type of turbine must always be pointed in the direction of the wind if it is to be used efficiently. Older and smaller wind turbines of the horizontal type can change direction via a simple wind vane, while large and more modern turbines have special motors in them which allow them to change direction as the wind shifts.

Vertical axis wind turbines

In contrast to horizontal axis wind turbines, vertical axis wind turbines, shortened to VAWTs, are designed with a vertical rotor shaft, a generator and gearbox, all of which are placed at the bottom of the turbine, and a uniquely shaped rotor blade that is designed to harvest the power of the wind no matter in which direction it is blowing.

One particular type of VAWT is the Darrieus wind turbine, which is designed to look like a modified egg beater. These turbines have very good efficiency but poor reliability due to the massive amount of torque which they exert on the frame. Furthermore, they also require a small generator to get them started.

Figure 10.11 shows some of the most common wind turbines and their respective efficiencies.

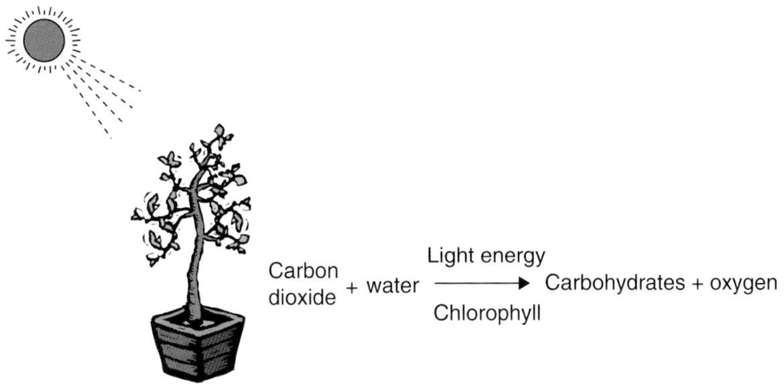

$$\text{Carbon dioxide} + \text{water} \xrightarrow[\text{Chlorophyll}]{\text{Light energy}} \text{Carbohydrates} + \text{oxygen}$$

Figure 10.12   Photosynthesis.

10.4   Biomass

10.4.1   Sources of biomass

The most common form of biomass is wood. For thousands of years people have burned wood for heating and cooking. Wood was the main source of energy of the world until the mid-1800s, and it continues to be a major source of energy in much of the developing world.

The discovery of coal and the Industrial Revolution in the United Kingdom changed all that; our energy became more diverse, and its uses spread to transport, machinery and leisure. The discovery of oil had even more impact on biomass, making it almost a redundant source of energy.

Biomass is considered a carbon-neutral fuel because during the growth of plants and trees, in the process of photosynthesis, they absorb carbon dioxide from the surrounding air and release oxygen to the atmosphere; in fact, the rate of production of oxygen during the growth of plants surpasses the CO_2 released during the combustion. Biomass is originally solar energy converted into chemical energy stored in organic matter.

Green plants invoke a process called photosynthesis whereby they absorb light energy from the sun and use the absorbed energy to synthesise compounds called carbohydrates (sugars, starches and cellulose), using carbon dioxide and water as the raw materials for the synthesis (Figure 10.12).

Photosynthesis is actually a two-stage process:

- A *light reaction*, in which the energy from absorbed sunlight is used to split water molecules into hydrogen and oxygen.
- A *dark reaction*, whereby hydrogen from the light reaction combines with carbon dioxide to form carbohydrates.

Biomass, by definition, can be found in the following sources:

- *Short-rotation coppice*. This incorporates stand-alone plots of land used to grow trees and plants for the purpose of being used as fuel, usually plants such as elephant grass and poplar trees.

Table 10.3   Comparison of biomass with other fuels.

Fuel	Net calorific value (MJ/kg)	Carbon content (%)	Direct carbon emission from combustion		Annual total CO$_2$ emissions to heat a typical house (20 000kWh/yr)		
			kg/GJ	kg/MWh	kg	kg saved compared with oil	kg saved compared with gas
Coal	29	75	26	94	9680	−2680	−4280
Oil	42	85	20	72	7000	0	−1600
Natural gas	38	73	19	69	5400	1600	0
LPG	46	82	18	64	6460	540	−1060
Electricity (UK grid)	–	–	35	125	10600	−3600	−5200
Electricity (large-scale wood chip combustion)	–	–	160	576	1160	5840	4240
Electricity (large-scale wood chip gasification)	–	–	80	286	500	6500	4900
Wood chips (25% MC) Fuel only	14	37.5	27	96	140	6860	5260
Wood pellets (10% MC starting from dry wood waste)	17	45	26	95	300	6700	5100
Grasses/straw (15% MC)	14.5	38	26	95	108 to 300	6892 to 6700	5292 to 5100
Biogas (60% CH$_4$ 40% CO$_2$)	30	56	19	67	–	–	–

Source: Oberweis, S. and Al-Shemmeri, T. T. (2008) Design of a 20kW Biomass Heat Generator: CFD Modelling and Dissociation, VDM Verlag.

- *Farm waste.* The straw of cereals and pulses, stalks and seed coats of oil seeds, stalks and sticks of fibre crops, pulp and waste from plantation crops, peelings, pulp and stalks of fruits and vegetables and other waste like sugarcane trash, rice husk, molasses, coconut shells, etc. come under this category.
- *Forest waste.* Logs, bark chips and leaves together constitute forest waste. Sawdust is the forest-based industry waste. Forest products are also used as a domestic fuel in many developing countries. Tree roots, bark and knots from a standard logging operation and broken debris generally are also considered forest waste.
- *Industrial waste.* This includes waste from paper mills and chemical mills, etc. For example, paper waste, plastic waste, textile waste, gas, oil, paraffins, cotton seeds and fibres, bagasse, etc. Waste from plastic and rubber has a good calorific value.
- *Municipal solid waste.* Generally, municipal solid waste refers to a mixture of domestic, small construction and demolition waste left out within a community. The composition of municipal solid waste is heterogenic; it typically contains 40% paper, 20% vegetables and the remaining 40% is made up of plastic, metal and glass.
- *Municipal sewage sludge.* The sludge contains organic matter and nutrients as the main constituents. These can be utilised for the production of methane through anaerobic digestion.
- *Animal waste.* Manure or dung is used as a fuel in China, India and Pakistan.

10.4.2   Combustion equation for biomass

Wood combustion processes are quite complex due to the nature of the fuel and its nonuniformity. For dry wood (zero moisture), the process is simplified by the following combustion equation when chemically correct oxygen is used:

$$C_{4.17}H_{6.5}O_{2.71} + 4.44\ O_2 \rightarrow 4.17\ CO_2 + 3.25\ H_2O \qquad [10.24]$$

The ultimate analysis of dry wood (that is, the ratio of carbon, hydrogen, oxygen and other elements) varies slightly from one species to another, but for most practical purposes, the above equation is adequate.

The combustion process described above releases about 10–20 MJ per kg of dry wood burnt; see Table 10.3.

The following equation is the result of fitting the data in Table 10.3 to evaluate the higher heating value (HHV) of biomass:

$$\text{HHV (in kJ/g)} = 0.3491\ C + 1.1783\ H - 0.1034\ O - 0.0211\ A + 0.1005\ S$$
$$- 0.0151\ N \qquad [10.25]$$

where C is the weight fraction of carbon; H of hydrogen; O of oxygen; A of ash; S of sulphur and N of nitrogen appearing in the ultimate analysis.

When burned, the chemical energy in biomass is released as heat. This could be used directly, as in a fireplace to heat a building, or to produce steam for

Table 10.4 Biomass technology.

Technology	Conversion process type	Major biomass feedstock	Energy or fuel produced
Direct combustion	Thermochemical	Wood, agricultural waste, municipal solid waste, residential fuels	Heat, steam, electricity
Gasification	Thermochemical	Wood, agricultural waste, municipal solid waste	Low or medium-producer gas
Pyrolysis	Thermochemical	Wood, agricultural waste, municipal solid waste	Synthetic fuel oil (biocrude), charcoal
Anaerobic digestion	Biochemical (anaerobic)	Animal manure, agricultural waste, landfills, wastewater	Medium gas (methane)
Ethanol production	Biochemical (aerobic)	Sugar or starch crops, wood waste, pulp, sludge, grass, straw	Ethanol
Biodiesel production	Chemical	Rapeseed, soy beans, waste vegetable oil, animal fats	Biodiesel
Methanol production	Thermochemical	Wood, agricultural waste, municipal solid waste	Methanol

making electricity. There are many different ways of using biomass for energy, and the form and properties of the biomass, together with the needs of the user, will determine which are the most appropriate. Table 10.4 shows the different technologies available for utilising biomass.

10.5   Hydraulic turbines

Hydroelectric power uses the kinetic energy of moving water to make electricity. Dams can be built to accumulate water and increase its height; the potential energy is then released, converted into kinetic energy and allowed to pass against a solid object, imparting its energy and hence making it rotate.

For hundreds of years, moving water was used to turn wooden wheels that were attached to grinding wheels to grind (or mill) flour or corn.

Today, hydraulic power is used to rotate hydraulic turbines, which are much more efficient than the old mills and are usually connected to electrical generators, designed to produce electricity (Figure 10.13).

10.5.1   Theory of hydraulic turbines

The principles concerned with converting the potential energy of water into useful power rely on three fundamentals: conservation of mass, conservation

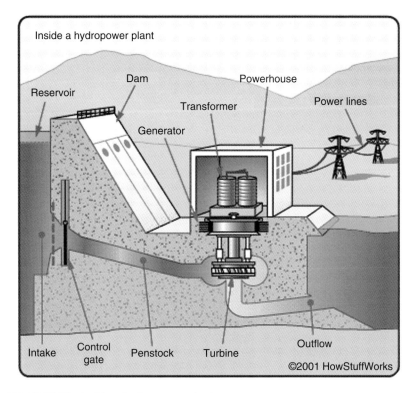

Figure 10.13 Inside a hydropower plant.

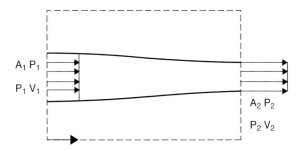

Figure 10.14 Conservation of mass of a fluid flowing in a duct/pipe.

of energy and conservation of momentum. It is therefore useful to discuss these before examining the operation of hydraulic turbines.

Conservation of mass

The *continuity equation* applies the principle of conservation of mass to fluid flow. Consider a fluid flowing through a fixed conduit with one inlet and one outlet, as shown in Figure 10.14.

If the flow is *steady*, i.e there is no accumulation of fluid within the control volume, the rate of fluid flow at entry must be equal to the rate of fluid flow at exit for mass conservation. If the flow's cross-sectional area is A (m²) and the fluid parcel travels a distance dL in time dt, then the volume flow rate ($\dot{V}$, m³/s) is given by:

$$\dot{V} = (A \, dL)/dt$$

but since dL/dt is the fluid velocity ($\bar{V}$, m/s), we can write:

$$\dot{V} = \bar{V} A$$

The mass flow rate ($\dot{m}$, kg/s) is given by the product of the density and the volume flow rate:

$$\dot{m} = \rho \dot{V} = \rho \bar{V} A$$

Between any two points within the control volume, the fluid mass flow rate can be shown to remain constant:

$$\dot{m}_1 = \dot{m}_2$$

or

$$\rho_1 \bar{V}_1 A_1 = \rho_2 \bar{V}_2 A_2 \qquad\qquad [10.26]$$

Conservation of energy

Conservation of energy necessitates that the total energy of the fluid remains constant; however, there can be transformation from one form to another.

There are three forms of nonthermal energy for a fluid at any given point:

- the *kinetic energy* due to the motion of the fluid;
- the *potential energy* due to the positional elevation above a datum;
- the *pressure energy*, due to the absolute pressure of the fluid at that point.

If all energy terms are written in the form of the *head* (potential energy), i.e. in metres of the fluid, then the conservation of energy principle requires that:

$$\left(\frac{p}{\rho g} + \frac{\bar{V}^2}{2g} + z \right)_1 = \left(\frac{p}{\rho g} + \frac{\bar{V}^2}{2g} + z \right)_2 \qquad\qquad [10.27]$$

This equation is known as the *Bernoulli equation* and is valid if the two points of interest (1 and 2) are very close to each other and there is no loss of energy.

In a real situation, the flow will suffer a loss of energy due to friction (h_L) and obstruction between stations 1 and 2, hence:

$$\left(\frac{p}{\rho g}+\frac{\bar{V}^2}{2g}+z\right)_1=\left(\frac{p}{\rho g}+\frac{\bar{V}^2}{2g}+z\right)_2+h_L \qquad [10.27a]$$

Flow regimes

In 1883, Osborne Reynolds demonstrated experimentally that, under laminar flow, the fluid streamlines remain parallel. This was shown with the aid of a dye filament injected into the flow which remained intact at low flow velocities in the tube. As the flow velocity was increased (via a control valve), a point was reached at which the dye filament at first began to oscillate then broke up so that the colour was diffused over the whole cross-section, indicating that particles of fluid no longer moved in an orderly manner but occupied different relative positions in successive cross-sections downstream.

Reynolds also found that it was not only the average pipe velocity ($\bar{V}$) which determined whether the flow was laminar or turbulent, but that the density (ρ) and viscosity (μ) of the fluid together with the pipe diameter (D) also determined the flow regime. He proposed that the criterion which determined the type of regime was the dimensionless group $\rho\bar{V}D/\mu$. This group has been called the *Reynolds number* (Re) as a tribute to his contribution to fluid mechanics.

$$\mathrm{Re}=\rho\bar{V}D/\mu \qquad [10.28]$$

With the Reynolds number, the flow can be distinguished into three regimes:

- laminar if Re < 2000;
- transitional if 2000 < Re < 4000;
- turbulent if Re > 4000.

Re = 2000, 4000 are the lower and upper critical values.
The Moody chart clearly demonstrates these flow regimes.

Conservation of momentum

Consider a duct of length L, cross-sectional area A_c and surface area A_s in which a fluid of density ρ is flowing at mean velocity $\bar{V}$.

The forces acting on a segment of the duct are that due to pressure difference and that due to friction at the walls in contact with the fluid.

If the acceleration of the fluid is zero, the net forces acting on the element must be zero, hence

$$(p_1-p_2)\,A_c=(f\,\rho\,\bar{V}^2/2)\,A_s$$

where f is known as the friction factor.

Let h_f denote the head lost (m) due to friction over the duct length L, i.e.:

$$p_1 - p_2 = \rho\, g\, h_f$$

Substituting, we get

$$h_f = f\,(A_s/A_c)\,\bar{V}^2/2g$$

For a pipe,

$$A_s/A_c = \pi DL/\pi D^2/4 = 4L/D$$

Hence

$$h_f = (4fL/D)\,\bar{V}^2/2g \qquad\qquad [10.29]$$

This is known as Darcy's formula.

The value of the friction factor (f) depends mainly on two parameters: the value of the Reynolds number and the surface roughness.

For *laminar* flow (i.e. Re < 2000),

$$f = \frac{16}{Re} \qquad\qquad [10.30]$$

For a *smooth* pipe with *turbulent* flow (i.e. Re > 4000),

$$f = \frac{0.079}{Re^{0.25}} \qquad\qquad [10.31]$$

The region where Re > 2000 and Re < 4000 is known as the *critical zone* and the value of the friction factor is certain.

In the turbulent zone, if the surface of the pipe is not perfectly smooth, then the value of the friction factor has to be determined from the *Moody diagram*. The *relative roughness* is the ratio of the average height of the surface projections on the inside of the pipe (k) to the pipe diameter (D). In common with the Reynolds number and the friction factor, this parameter is dimensionless.

Flow obstruction losses

When a pipe changes direction, changes diameter or has a valve or other fittings, there will be a loss of energy due to the disturbance in flow. This loss of ener gy (h_o) is usually expressed by:

$$h_o = K\frac{\bar{V}^2}{2g} \qquad\qquad [10.32]$$

Table 10.5 Typical obstruction losses in flow systems.

Obstruction		K
Tank exit		0.5
Tank entry		1.0
Smooth bend		0.30
Mitre bend		1.1
Mitre bend with guide vanes		0.2
90° elbow		0.9
45° elbow		0.42
Standard T		1.8
Return bend		2.2
Strainer		2.0
Globe valve, wide open		10.0
Angle valve, wide open		5.0
Gate valve, wide open		0.19
¾ open		1.15
½ open		5.6
¼ open		24.0
Sudden enlargement		0.10
Conical enlargement: 6°		0.13
(total included angle) 10°		0.16
15°		0.30
25°		0.55
Sudden contractions:		
area ratio	0.2	0.41
(A_2/A_1)	0.4	0.30
	0.6	0.18
	0.8	0.06

where $\bar{V}$ is the mean velocity at entry to the fitting and K is an empirically determined factor; some examples are listed in Table 10.5.

10.5.2 Fluid power

The fluid power available at a given point is given by:

$$P = \rho g \, \dot{V} h_{tot} \qquad [10.33]$$

For a *pump*, h_{tot} represents the head required to overcome pipe friction (h_f), obstruction losses (h_o) and to raise the fluid to any elevation required (h_z) (see Figure 10.15), i.e.:

$$h_{tot} = h_z + h_f + h_o \qquad [10.34]$$

Note that if the delivery tank operates at a pressure in excess of that of the supply tank, an additional term (h_p) must be added to the required head equation, as this pressure rise must also be supplied by the pump.

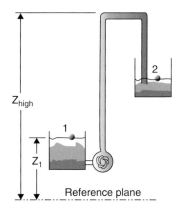

Figure 10.15 Fluid power system.

If the pump efficiency, η_p, is introduced, the actual pump head requirement is:

$$P = \rho g \, \dot{V} h_{tot}/\eta_p \qquad\qquad\qquad\qquad [10.33a]$$

For a *turbine* with efficiency η_t, the power output is given by:

$$P = \rho g \, \dot{V} h_{tot}/\eta_t \qquad\qquad\qquad\qquad [10.33b]$$

where

$$h_{tot} = h_z - (h_f + h_o) \qquad\qquad\qquad\qquad [10.34a]$$

10.5.3 Classification of hydraulic turbines

Hydraulic turbines have a row of blades fitted to a rotating shaft or a rotating plate. Flowing liquid, mostly water, passes through the hydraulic turbine, striking the blades and making the shaft rotate. While flowing through the hydraulic turbine, the velocity and pressure of the liquid reduce, resulting in the development of torque and rotation of the turbine shaft. There are different forms of hydraulic turbine (Figure 10.16) and their use depends on the operational requirements. Based on the flow path of the liquid, hydraulic turbines can be categorised into three types:

- *Axial flow hydraulic turbines*. This category has the flow path of the liquid mainly parallel to the axis of rotation. Kaplan turbines have the liquid flowing mainly in the axial direction.
- *Radial flow hydraulic turbines*. Such turbines have the liquid flowing mainly in a plane perpendicular to the axis of rotation.
- *Mixed flow hydraulic turbines*. For most of the hydraulic turbines used there is a significant component of both axial and radial flow. Such types are called mixed flow turbines. The Francis turbine is an example of a mixed flow type; here, water enters in a radial direction and exits in the axial direction.

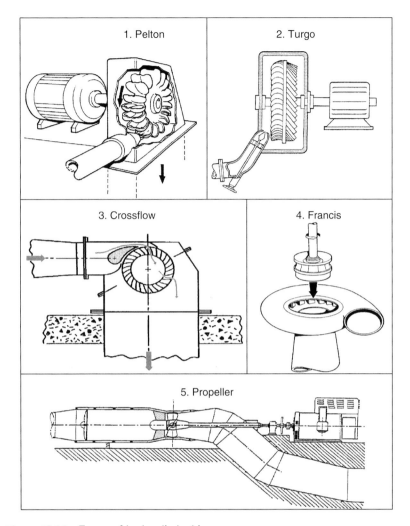

Figure 10.16   Types of hydraulic turbine.

None of the hydraulic turbines have purely axial flow or purely radial flow. There is always a component of radial flow in axial flow turbines and of axial flow in radial flow turbines.

Classification based on pressure change

One more important criterion to consider when classifying hydraulic turbines is whether or not the pressure of liquid changes while it is flowing through the rotor of the turbine. Based on the pressure change, hydraulic turbines can be categorised into two types:

● *Impulse turbines*. The pressure of liquid does not change while flowing through the rotor of the machine. In impulse turbines, pressure change occurs only in the nozzles of the machine. One example of an impulse turbine is the Pelton wheel.

- *Reaction turbines*. The pressure of liquid changes while flowing through the rotor of the machine. The change in fluid velocity and reduction in its pressure causes a reaction on the turbine blades; hence the probable derivation of the name reaction turbine. Francis and Kaplan turbines fall into this category.

Pelton wheel or turbine

The Pelton wheel is used where a small flow of water is available with a 'large head'. It resembles the waterwheels used at water mills in the past. The Pelton wheel has small 'buckets' all around its rim. Water from the dam is fed through nozzles at very high speed, hitting the buckets and pushing the wheel around.

The Turgo turbine

This is an impulse turbine that is similar to the Pelton wheel but can handle higher flow rates, although it is somewhat less efficient. The runner (wheel) of a Turgo turbine is different to the Pelton wheel; its incoming jet of water strikes at an angle of about 20° and exits on the other side of the runner, therefore the incoming and outgoing jets don't interfere (as they do in the case of the Pelton wheel). The Turgo runner is effectively a Pelton runner split down the middle, therefore it can generate the same power as a Pelton wheel with twice the diameter and thus it has twice the specific speed of the Pelton wheel and can handle larger flow rates than a similar-sized Pelton wheel.

The Francis turbine

The Francis turbine is used where a large flow and a high or medium head of water is involved.

This turbine is also similar to a waterwheel, as it looks like a spinning wheel with fixed blades in between two rims. The wheel is called a 'runner'. A circle of guide vanes surround the runner and control the amount of water driving it. Water is fed to the runner from all sides by these vanes, causing it to spin.

The Kaplan turbine

Propeller-type turbines are designed to operate where a small head of water is involved. These turbines resemble ships' propellers. However, with Kaplan turbines the angle (or pitch) of the blades can be altered to suit the water flow.

The variable-pitch feature permits the machine to operate efficiently over a range of heads, to allow for the seasonal variation of water levels in a dam.

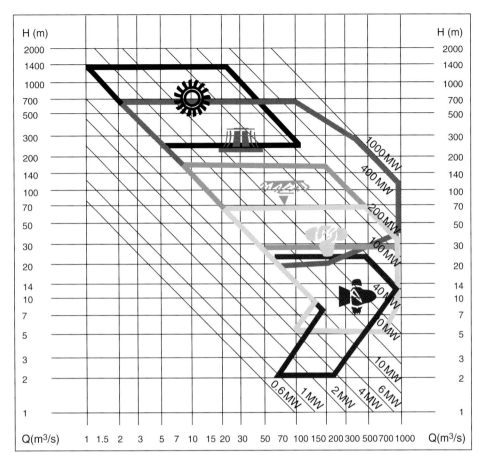

Figure 10.17   Characteristics of hydraulic turbines.

10.5.4   Design and selection of hydraulic turbines

Turbine selection is based mostly on the available water head, and less so on the available flow rate. In general, impulse turbines are used for high-head sites and reaction turbines are used for low-head sites. Kaplan turbines with adjustable blade pitch are well adapted to wide ranges of flow or head conditions, since their peak efficiency can be achieved over a wide range of conditions. Figure 10.17 demonstrates the relationship between P, H and Q for different hydraulic turbines:

$$\text{Hydraulic power} = \text{efficiency of hydraulic turbine} \times \text{water input power}$$

10.5.5   Relationship between specific speed and type of hydraulic turbine

The specific speed of a turbine is a dimensionless parameter defined as the speed of an ideal, geometrically similar turbine, which yields one unit of power for one unit of head:

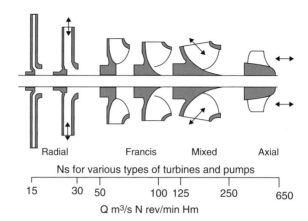

Radial   Francis   Mixed   Axial

Ns for various types of turbines and pumps

15   30 50   100 125   250   650

Q m³/s N rev/min Hm

Figure 10.18 Specific speeds for hydraulic turbines.

$$n_s = n\sqrt{P}/H^{5/4}$$ [10.35]

where:

n is the wheel speed in rpm
P is the power in kW
h is the water head in metres

Various specific values for n_s for different types of turbine are given in Figure 10.18.

10.6   Worked examples

Worked example 10.1

Calculate the area of PV panels (roof-mounted system) required to produce 10% of the electricity requirement for the United Kingdom. Assume that:

- Annual radiation per square metre of tilted surface = 1065 kWh/m²/yr (figure for Kew, London with a panel tilt of 30°).
- Number of homes with roofs suitable for PV installations = 10 million. (This represents 50% of the UK housing stock.)
- Assumed efficiency of a PV panel = 15%.
- Each home consumes 160 kWh.
- Total UK energy consumption is 300 TWh.

Solution:

The net output per annum per m² of PV panel is 1065 × 15/100 = 160 kWh.

10 million homes $\times$ 160 kWh $= 10^7 \times$ 160 kWh $= 1.6$ TWh.

To provide 30 TWh, each of the 10 million PV panel arrays would need to have an active area of approximately 19 m² (30/1.6 = 18.75).
The total area of PV panels is:

19 m² $\times 10^7 = 1.9 \times 10^8$ m²

One hectare is equivalent to 10 000 m².
The total area of PV panels in hectares is:

$1.9 \times 10^8 / 1.0 \times 10^4 = 19\,000$

This is a total area of land about half the size of the Isle of Wight.

Worked example 10.2

Worked example 10.2, Table 1 summarises the data for a wind turbine to be installed. Determine the annual power output for this turbine on this particular site.

Worked example 10.2, Table 1

Set	Wind speed (m/sec)	Turbine output (kW)	No of hours per year at given wind speed
(a)	(b)	(c)	(d)
1	4	2	1100
2	5	4	1100
3	6	6	1000
4	7	8	900
5	8	10	800
6	9	10	600
7	10	10	400
8	11	10	300
9	12	10	200
10	13	10	100

Solution:

First calculate the power output for each speed group, then add them up, as shown in Worked example 10.2, Table 2. The annual output is therefore calculated as 43.8 MWh per year.

Worked example 10.2, Table 2

Set	Wind speed (m/sec)	Turbine output (kW)	No of hours per year at given wind speed	Power output kWhr (c × d)
(a)	(b)	(c)	(d)	(e)
1	4	2	1100	2200
2	5	4	1100	4400
3	6	6	1000	6000
4	7	8	900	7200
5	8	10	800	8000
6	9	10	600	6000
7	10	10	400	4000
8	11	10	300	3000
9	12	10	200	2000
10	13	10	100	1000
				43 800

Worked example 10.3

Worked example 10.3, Figure 1 shows what happens to the sun's radiation when it falls on a panel of solar cells.

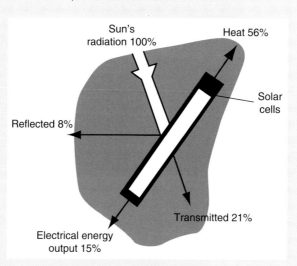

Worked example 10.3, Figure 1

Each square metre of the solar panel receives 500 W. The area of the panel is 5 m². How much electricity will be produced by these panels in two hours?

Solution:

Solar energy received = (solar intensity × area of cells) × time
$$= (500 \times 5) \times 2$$
$$= 5000 \, Wh$$

Multiply this by 3.6 to get 18 000 kJ, which is 18 MJ.
   The electricity generated is 15% of this value, i.e. 2.7 MJ.

Worked example 10.4

Estimate the PV array sizes necessary to generate 10kWh of usable energy with an average daily insolation of $2 \, kWh/m^2/day$. Assume that the efficiency of PV cells is 15%.

Solution:

To get 10kWh from a daily insolation of $2 \, kWh/m^2$, an area of $5 \, m^2$ is needed if the PV cells are 100% efficient.
   Using a photovoltaic array with a conversion efficiency of 15%, the area of the array will be:

$$5/(0.15) = 33.3 \, m^2$$

Note that the PV array output is DC electrical power. To provide AC power, there would be further electrical losses of 10% to 20% in the voltage regulator, inverter and control circuits.
   Assuming 20% electrical system losses, a fixed PV array with an area of around $40 \, m^2$ would be required to provide 10 kWh of AC power per day.

Worked example 10.5

A site receives $1000 \, W/m^2$ of solar radiation in July. Assume that the solar panels are 10% efficient and that the average sunny day is 5 hours. How many square metres would be required to generate 10 kWh of electricity? If a typical solar panel is $5 \, m^2$, how many panels will be needed?

Solution:

   Energy output = incident energy × panel efficiency

Hence, each square metre gives you $1000 \times 0.1 = 100$ Watts.

In five hours you would get:

$5 \times 100 = 500$ Watt-hours or about 0.5 kWh per square metre

In order to get 10 kWh,

Area needed $= 10/0.5 = 20$ square metres of collecting area

The number of panels needed $= 20/5 = 4$ panels.

Worked example 10.6

The average wind speed of 10 mph yields 100 W/m². Determine the power produced by a windmill when the wind speed is 40 mph.

Solution:

The change in wind speed means that the wind blows at the ratio of:

$40/10 = 4$ times

Since the power is proportional to wind speed to the power of 3, an increase in wind velocity of 4 times implies that the power will be increased by a factor of:

$4^3 = 64$

Therefore, if a 10 mph wind gives you 100 Watts, then a 40 mph wind gives you 64 times more power

$= 64 \times 100 = 6400$ Watts
$= 6.4$ kilowatts.

Worked example 10.7

An individual photovoltaic cell with an active area of 0.04 m² operates at solar-to electrical efficiency of 6 when the incident solar intensity is 500 W/m².

(a) Calculate the power output from the cell.
(b) Assuming that it is maintaining a voltage of 0.5 Volts, determine the current delivered.
(c) How many cells are required to produce an output of 14V, 7.2 Amps?

Solution:

(a) The power output

E = efficiency of the cell × A × I
   = 0.06 × 0.04 × 500
   = 1.2 W

(b) The electrical current in the circuit:

$I = E/V$
   = 1.2/0.5
   = 2.4 Amps

(c) To get 14 Volts, 28 cells are needed, each at 0.5 V. To get 7.2 Amps, we need three rows of 2.4 Amps. Hence, the total number of cells = 28 × 3 = 84.

Worked example 10.8

The specifications set out in Worked example 10.8, Table 1 for two HAWTs are supplied by the manufacturers.

Worked example 10.8, Table 1

Item	Turbine A	Turbine B
Rotor diameter	25 m	28 m
Power coefficient	38	35
Gearbox efficiency	90	88
Generator efficiency	98	95
Capital cost	£99 000	£103 000
Maintenance cost/year	£4000	£4000

(a) Draw up a table for the performance of each turbine for wind speeds of 4–12 m/s in intervals of 2 m/s.
(b) Assume the site wind availability to be 2000 hours per year, and average wind speed to be 6 m/s. Select the wind turbine which will be the most economical. Assume the life expectancy for each to be 20 years, and the unit cost of power to be 6 pence per kWh to remain constant.

Solution:

(a) The shaft power of a wind rotor is given by:

$$P = C_p \times 1/2 \times \rho \times A \times V^3 \times \eta_{gb} \times \eta_{gen}$$

The relevant data are given in Worked example 10.8, Table 2.

Worked example 10.8, Table 2

Wind speed m/s	Turbine A Power (kW) = 98.713V³	Turbine B Power (kW) = 108.101V³
4	6.317	6.918
6	21.322	23.350
8	50.541	55.348
10	98.713	108.101
12	170.576	186.799

(b) At 6m/s, Turbine B produces more power.

Annual difference = $(23.350 - 21.322) \times 2000 = 4056$ kWh
Cost difference over 20 years = $4056 \times 0.06 \times 20 = £4867$

Hence, Turbine B is chosen even though it is £4000 more expensive to buy than A. There is still a net saving of £867.

Worked example 10.9

Water supplied from a local reservoir at 0.28 m³/s under a head of 72 m is to be used to produce hydro-electricity. What kind of turbine is best suited to this application? Assume a hydraulic efficiency of 80% and a rotational speed of 600 rpm.

Solution:

Power output = efficiency × water input power
$P = \eta_h \times \rho \times g \times H \times Q$
$\quad = 0.8 \times 1000 \times 9.81 \times 72 \times 0.28$
$\quad = 158$ kW

$$N_s = \frac{N\sqrt{P}}{H^{1.25}} = \frac{600\sqrt{158}}{72^{1.25}} = 8.65$$

It would therefore be necessary to use a Turgo turbine.
However, it might be possible to use a Pelton wheel with two jets:

$$\therefore \ N_s \ per \ Jet = \frac{9.83}{\sqrt{2}} = 6.95$$

Try a Pelton wheel with four jets:

$$\therefore \ N_s \ per \ Jet = \frac{9.83}{\sqrt{2}} = 4.92$$

This would be a practical proposition but would result in some loss of efficiency due to interference between the jets. Another alternative would be to have two wheels on the same shaft with two jets per wheel.

Worked example 10.10

Calculate the specific speeds and suggest the most suitable type of hydraulic turbine for some sites in Galloway using the data given in Worked example 10.10, Table 1.

Worked example 10.10, Table 1

H	N	P
m	rpm	kW
11	300	2000
46	250	12 000
20	214.3	6000
20	214.3	7000
116	428.6	12 000
32	214.3	11 000
400	300	15 000

Solution:

Using the definition, the specific speed is calculated as:

$$N_s = \frac{N\sqrt{P}}{H^{1.25}}$$

And, depending on the value of this specific speed, the type of turbine is selected according to Figure 10.18. Hence, the solution can be shown as in Worked example 10.10, Table 2.

Worked example 10.10, Table 2

H	N	P	N_s	Type
m	rpm	kW		
11	300	2000	669.72	Kaplan
46	250	12 000	228.60	Francis
20	214.3	6000	392.47	Kaplan
20	214.3	7000	423.92	Kaplan
116	428.6	12 000	123.33	Francis
32	214.3	11 000	295.31	Francis
400	300	15 000	20.54	Pelton

Worked example 10.11

Water is routed from a reservoir to a proposed hydroelectric scheme at the rate of 30 m³/s with a head of 4 m.

(a) Calculate the hydraulic power available.
(b) Two generators are available and you are required to select the correct hydraulic turbine for each option, justifying your selection:
  • option 1 - with a rotational speed of 60 rpm;
  • option 2 - with a rotational speed of 120 rpm.
  • Both turbines have a hydraulic efficiency of 90%.

Solution:

(a)

Total hydraulic power = efficiency × water input power

$$P = \eta_h \times \rho \times g \times H \times Q$$
$$= 0.9 \times 1000 \times 9.81 \times 4 \times 30 = 1060 \, kW$$

(b) Since both options have a hydraulic efficiency of 90%, this figure applies to both.

Option 1, $N = 60$ rpm

$$N_s = \frac{N\sqrt{P}}{H^{1.25}} = \frac{60 \times \sqrt{1060}}{4^{1.25}} = 345 \, rpm$$

Hence, a Francis turbine is suitable.

Option 2, $N = 120$ rpm

$$N_s = \frac{N\sqrt{P}}{H^{1.25}} = \frac{120 \times \sqrt{1060}}{4^{1.25}} = 690 \, rpm$$

Hence, a Kaplan turbine is suitable.

10.7 Tutorial problems

10.1 A solar cell with an active surface of 100 mm by 100 mm is exposed to a light intensity of 1000 Watts/m².

(a) Calculate the power incident on the cell.
(b) When the cell is operating at 10% efficiency, what electrical power is it supplying?
(c) If the cell maintains a voltage of 0.5 V, what current is it delivering?

Ans. (10 W, 1 W, 2 Amps)

10.2 A site receives 600 Watts/m² of solar radiation in July. Assume that solar panels are 10% efficient and that the average sunny day is 6 hours.

(a) How many square metres would be required to generate 10 kWh of electricity?
(b) If a typical solar panel is 1.2 m², how many panels will be needed?

Ans. (27.8 m², 24 panels)

10.3 Calculate the power delivered by the wind, in W/m², for wind speed over a range from 1 to 20 m/s. (Assume the density of air at normal pressure to be 1.23 kg/m³.)
Ans. (1.6 kW/m² at 20 m/s)

10.4 The rotor diameter of a two-bladed HAWT is 7.0 metres. At its rated wind speed of 12.1 m/s, the power output of the turbine is 15 kW.

(a) Find the efficiency of the turbine at this wind speed. (The density of air is 1.23 kg/m³.)
(b) Calculate the tip-speed ratio if the blades are rotating at 240 rpm.

Ans. (0.357, 7.27)

10.5 A two-bladed horizontal axis wind generator has a rotor diameter of 20 metres.

(a) Operating at a wind speed of 8.1 m/s, the rotor extracts 35% of the energy of the wind. If the efficiency of the generator which it drives is 85%, find the electrical power output in kilowatts. (The density of air is 1.23 kg/m³.)
(b) The output power of a turbine is equal to the torque exerted by the blades multiplied by the angular velocity (which is

approximately 0.1 times the rpm). If the above turbine is rotating at 60 rpm, what is the torque?

Ans. (30.5 kW, 4.8 kNm)

10.6 The Dinorwig pumped storage system in Wales can store 7.2 million cubic metres of water at an average height of 500 metres above its lower reservoir.

(a) Calculate the maximum energy stored.
(b) The water is pumped up overnight by six turbines operating as pumps, each consuming 281 MW. If the energy conversion efficiency of the system when pumping is 94%, estimate the time it takes to pump the full 7.2 million cubic metres of water.
(c) The maximum output of each of the six generators of the system is 306 MW. If this output is produced when the height difference between the upper and lower reservoirs is 510 metres and the total flow rate to all six turbines is 390 m³/s, what is the overall energy conversion efficiency?
(d) How many kilowatt-hours are 'lost' in a complete cycle of storage and generation if the above results represent the average efficiencies of pumping and generating?
(e) The Dinorwig turbines rotate at 500 rpm. Calculate the specific speed under the above generating conditions. What type of turbine would you expect them to be?

Ans. (9.8 × 10⁶ MWh, 5.5 h, 94%, 0.589 × 10⁶ MWh, 112, mixed turbine)

10.7

(a) Find the volume flow rate required to deliver a power of 2.25 MW with each of the following effective heads: 50 m, 250 m and 1000 m.
(b) Assuming no energy losses, calculate the speed at which water emerges from a jet after falling through each of the above heads.
(c) Use the results of (a) and (b) to find the jet diameter required to deliver the 2.25 MW from each head.

Ans. ((a): 4.587, 0.917, 0.229 m³/s; (b): 31.3, 70.0, 140 m/s; (c): 0.432, 0.129, 0.046 m)

Appendix

Case Study: Energy audit for a school

In this section, we conduct an energy audit on a local school in accordance with the procedure outlined earlier. Permission was sought to undertake the work, and this helped to ensure that all the necessary data were available.

The school under consideration is Springhead Primary School, Talke Pit, Newcastle (Figure A.1), which is located in a built-up area of Staffordshire, UK.

The building has:

9 teaching classrooms
100 students and 20 teaching staff
16 printers
3 photocopiers
41 desktop computers
2 ovens
2 fridges
5 toasters
1 kettle
16 hi-fis
4 projectors
2 TVs

A.1 Energy consumption details

Energy consumption over the one-year period from January to December for both gas and electricity can be seen in Table A.1. Also included is the annual average degree-day data for the Midlands.

A.2 Degree-day method

The analysis in this section follows the method outlined in Section 6.4.

Energy Audits: A Workbook for Energy Management in Buildings, First Edition.
Tarik Al-Shemmeri.
© 2011 Blackwell Publishing Ltd. Published 2011 by Blackwell Publishing Ltd.

Figure A.1   Springhead Primary School.

Table A.1   Total (kWh) of gas and electricity used for the school year 2008.

Month	Gas (Therm)	Degree -day	Gas (kWh)	Electricity (kWh)
January	796.87	376	23 212.8231	976.2
February	1593.73	359	46 425.3549	1952.4
March	1195.3	322	34 819.089	2043.5
April	413.3	243	12 039.429	1311.33333
May	826.7	162	24 081.771	2043.5
June	620	83	18 060.6	1967
July	138.2	44	4025.766	1323.78
August	276.4	48	8051.532	2647.56
September	207.3	90	6038.649	1985.67
October	524.4	178	15 275.772	2634.666
November	393.3	275	11 456.829	1317.3333
December	262.2	343	7637.886	1976
Total		2523	211 125.501	22 178.94263
Total gas and electric				233 304.4436

Using Table A.1, evaluation of future monitoring of the building's energy consumption has been represented on a graph (Figure A.2) detailing the regression equation and base energy value.

Figure A.2 shows the consumption of energy for each month of a 12-month calendar.

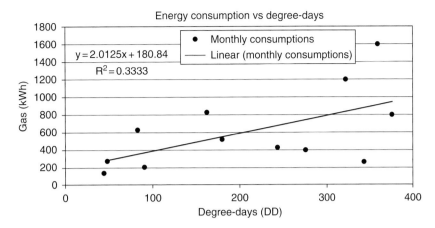

Figure A.2   Plot of energy consumption vs degree-days.

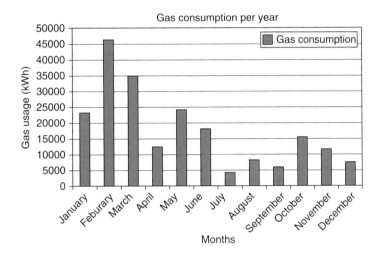

Figure A.3   Annual gas consumption.

Further analyses were also carried out in order to investigate the variation in electricity and gas usage over the course of one year, and these are shown in Figures A.3 and A.4. As would be expected, these figures show that the values of both electricity and gas consumption become higher during the winter period as it gets dark quicker and is cooler than the summer.

A.3   NPI method

The NPI method is one of the standard ways of reporting the results of an energy audit carried out in buildings, and this report will be using the format and standards. (This method was outlined in Section 6.2.)

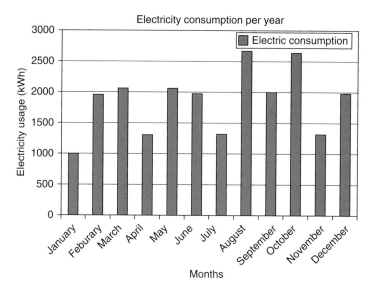

Figure A.4   Annual electricity consumption.

Some of the details are shown in Table A.2, but the analysis is summarised in the standard form for calculating NPI, as used in Chapter 2.

A.4   Energy-saving measures

In order to complete the audit, and because the building was found to be Fair according to the NPI analysis in Table A.2, a number of modifications will be conducted and investment made. In Table A.3, the original building's heating load is calculated.

Several cost-effective measures could improve the energy efficiency of the school and these will be considered next.

A.4.1   Improving insulation within the building

By improving the thermal insulation of the building's walls and roofs, heat loss can be reduced significantly and this will also reduce the hours of space heating, which, in turn, will save money on the annual energy bill.

Table A.4 shows the potential energy reduction possible if the walls and roof structure are improved such that the *U* values are 0.6 and 0.35 respectively. It can be seen that the new net heat transfer from the building is lower (26 kW) and this is about 61% of the actual net heat transferred (43 kW) calculated in Table A.3.

The NPI is recalculated and the results are shown in Table A.5. As can be seen, the upgrade has resulted in good news; the NPI verdict is GOOD.

Table A.2   NPI calculation form.

1. Convert your energy use into kWh units

Add your quarterly or monthly use over one year for each fuel and enter below

Natural gas	Therms × 29.31	kWh
	= 211 125.5 kWh	
Electricity	kWh × 1 = 22 178.9	kWh
Total energy use for the year	= 233 304 kWh	kWh	A

2. Find your space-heating energy use

If you can identify any of the fuels above used *only* for space heating, enter the total energy use in kWh

Add these to give total   kWh   B

If you cannot identify these, choose one of the following factors applied to the total energy used.

Annual space-heating energy	A × 0.75 = 0.75 × 211	kWh	C
	125.5 = 158 344		
	B or C =	kWh	D
Annual non-space-heating energy	A − D = 74 960	kWh	E

3. Adjust the space-heating energy to account for weather

Find the degree-days for the energy data year = 2523   F

The weather correction factor = 2462/2523 = 0.9758

Adjust the space-heating energy to standard conditions = D × G =   G

= 158 344 × 0.9758 = 154 515   kWh   H

Adjust the space-heating energy to account for exposure

Obtain the exposure factor from this chapter to suit the location   J
of the building = 1

Adjusted space-heating energy = H × J = 154 515   kWh   K

Find normalised annual energy use = E + K = 74 960 + 154 515   kWh
= 229 476   L

Correct for hours of use of building

Obtain standard hours of use from this chapter = 1400   M

Calculate the annual hours of use for your building = 195 days × 7 h   N
= 1365 h/year

Hours of use factor M/N = 1400/1365 = 1.026   P

Annual energy use for standard hours = P × L = 1.026 × 229 476   kWh   Q
= 235 360

Find floor area (or pool area) = 1073 m^2 R

Find the Normalised Performance Indicator (NPI) = Q/R kWh/m^2
= 235 360/1073
= 219

Compare NPI with yardsticks	FAIR
(Good, Fair or Poor)(180–239)	

Table A.3 Summarised heat load calculations.

Element	U value (W/m²K)	Area (m²)	Temperature difference (°C)	Heat loss (W)
Doors	3	28	6.25	525
Windows	3.5	267	6.25	5840.6
Walls	2.3	332.5	6.25	4779.7
Roof	2.2	1205	6.25	16 568.7
Floor	1.47	1073	6.25	9858.2
Fabric heat loss = Total (doors, windows, walls, roof and floor) =				37 572
Ventilation heat loss =				5818.7
Heat gain from occupants = 120 × 100 W =				12 000
Heat gain from lights = 49 × 38 W (with electronic ballast) =				1862
Heat gain from machines (TVs, printers, desktops, etc.) =				26 950
Net heat transfer from building =				43 kW

Table A.4 Improved heat loss calculations.

Element	U value (W/m²K)	Area (m²)	Temperature difference (°C)	Heat loss (W)
Doors	3	28	6.25	525
Windows	3.5	267	6.25	5840.6
Walls	0.6	332.5	6.25	1246.9
Roof	0.35	1205	6.25	2635.9
Floor	1.47	1073	6.25	9858.2
Fabric heat loss = Total (doors, windows, walls, roof and floor) =				20 106.6
Ventilation heat loss =				5818.7
Heat gain from occupants =				12 000
Heat gain from lights =				1862
Heat gain from machines (TVs, printers, desktops, etc.) =				26 950
Net heat transfer from buildings =				26 kW

Table A.5   NPI form, recalculated.

1. Convert your energy use into kWh units

Add your quarterly or monthly use over one year for each fuel and enter below

Natural gas	Therms × 29.31 = 128 786 kWh	kWh
	Cubic ft × 0.303 =	
Electricity	kWh × 1 = 22 178.9	kWh
Total energy use for the year	= 150 965 kWh	kWh A

2. Find your space-heating energy use

If you can identify any of the fuels above used *only* for space heating, enter the total energy use in kWh

Add these to give total   kWh   B

If you cannot identify these, choose one of the following factors applied to the total energy used.

Annual space-heating energy	A × 0.75 = 0.75 × 128 786	kWh C
	= 96 589	
	B or C = 96 589	kWh D
Annual non-space-heating energy	A−D = 54 375	kWh E

3. Adjust the space-heating energy to account for weather

Find the degree-days for the energy data year = 2523   F

The weather correction factor = 2462/2523 = 0.9758   G

Adjust the space-heating energy to standard conditions = D × G =

= 96 589 × 0.9758 = 94 251   kWh   H

Adjust the space-heating energy to account for exposure

Obtain the exposure factor from this chapter to suit the location of the building = 1   J

Adjusted space-heating energy = H × J = 94 251   kWh   K

Find normalised annual energy use = E + K = 54 375 + 94 251   kWh

= 148 626   L

Correct for hours of use of building

Obtain standard hours of use from this chapter = 1400   M

Calculate the annual hours of use for your building = 195 days ×   N
7h = 1365 h/year

Hours of use factor M/N = 1400/1365 = 1.026   P

Annual energy use for standard hours = P × L = 1.026 × 148 626   kWh   Q
= 152 490

Find floor area (or pool area) = 1073   m²   R

Find the Normalised Performance Indicator (NPI) = Q/R =   kWh/m²
= 152 490/1073 = 142

Compare NPI with yardsticks   Good

(Good, Fair or Poor)(180−239)

Calculating the savings

Actual original space-heating energy/year = 158 344.125 kWh
New space-heating energy/year = 96 589.9 kWh
Savings in energy/year = 158 344.125 − 96 589.9 = 61 754.2 kWh
Since the annual space-heating energy saving has been calculated, the cost savings in the annual energy bill can also be evaluated as follows:

Actual cost for space-heating energy /year

$$= 158344.125 \times \frac{1.7764 \text{ p/kWh}}{100} = £2812.83$$

New space-heating energy/year (improved insulation)

$$= 96589.9 \times \frac{1.7764 \text{p/kWh}}{100} = £1715.80$$

Savings in space-heating energy/year = 61 754.2 kWh

$$\text{Cost savings in energy bill/year} = 61754.2 \times \frac{1.7764 \text{p/kWh}}{100} = £1097$$

Pay-back period

On the basis of an estimate provided by builders, upgrading the *U* value of the walls and roof by introducing loft insulation and injecting cavity wall insulation would cost £6000; the pay-back period is estimated to be just under six years:

$$PBP = \frac{6000}{1097} = 5.47 \text{ years}$$

Discussion

The energy audit review showed that the school has an NPI rating of 219, which, according to the rating range (180–239), classifies Springhead's overall energy efficiency as fair. However, this can be improved by upgrading the roof and wall insulation to current UK insulation standards. (The lower the *U* value, the better the insulation.) The school qualifies for a 50% grant from the council to carry out such an upgrade and, as such, the pay-back period would be under three years.

A.4.2 Other options

Other energy-saving projects for the school to consider would include the following:

- *Upgrading the lights.* Improving the electrical lighting loads within the building would save money on energy bills and reduce the heat gain, which might contribute to summer peak load.

- *Light switching control.* Some of the electrical lighting loads could be reduced by using effective switches such as automatic photocell sensors or manual switches. Under bright enough conditions, the artificial lights would be switched off and replaced by natural daylight, which is free, thus reducing the carbon footprint.
- *Use of skylights.* It would also be possible to replace some of the artificial lighting with skylights, thus reducing electricity consumption.

A.5 Recommendations

It is our recommendation that the improvements suggested to the building's insulation are implemented, and that an investigation into the cost of materials and other strategies is carried out to further quantify the savings possible.

We also recommend the other saving options mentioned, i.e. light bulbs should be changed to more energy-efficient ones and natural daylight should be looked into.

A.6 Conclusion

This energy audit has been successful in improving the environmental impact, financial and operational costs of the school. This will go a long way to improving the productivity of the occupants.

As the world temperature is rising through our actions, it is important that we all reduce our energy consumption and prevent global warming from knocking even harder at the Earth's door.

Index

Keep up with critical fields

Would you like to receive up-to-date information on our books, journals and databases in the areas that interest you, direct to your mailbox?

Join the **Wiley e-mail service** - a convenient way to receive updates and exclusive discount offers on products from us.

Simply visit **www.wiley.com/email** and register online

We won't bombard you with emails and we'll only email you with information that's relevant to you. We will ALWAYS respect your e-mail privacy and NEVER sell, rent, or exchange your e-mail address to any outside company. Full details on our privacy policy can be found online.